STATISTICAL SIMULATION

Power Method
Polynomials and Other
Transformations

STATISTICAL SIMULATION

Power Method Polynomials and Other Transformations

Todd C. Headrick

Southern Illinois University

Carbondale, U.S.A.

CRC Press
Taylor & Francis Group
Boca Raton   London   New York

CRC Press is an imprint of the
Taylor & Francis Group an **informa** business

A CHAPMAN & HALL BOOK

Chapman & Hall/CRC
Taylor & Francis Group
6000 Broken Sound Parkway NW, Suite 300
Boca Raton, FL 33487-2742

First issued in paperback 2017

© 2010 by Taylor and Francis Group, LLC
Chapman & Hall/CRC is an imprint of Taylor & Francis Group, an Informa business

No claim to original U.S. Government works

ISBN 13: 978-1-138-11628-3 (pbk)
ISBN 13: 978-1-4200-6490-2 (hbk)

Library of Congress Cataloging-in-Publication Data

Headrick, Todd C.
 Statistical simulation : power method polynomials and other transformations / Todd C. Headrick.
     p. cm.
 Includes bibliographical references and index.
 ISBN 978-1-4200-6490-2 (hardcover : alk. paper)
 1. Monte Carlo method. 2. Distribution (Probability theory) I. Title.

QA298.H43 2010
518'.282--dc22   2009042286

Visit the Taylor & Francis Web site at
http://www.taylorandfrancis.com

and the CRC Press Web site at
http://www.crcpress.com

Contents

Preface

Monte Carlo and simulation techniques have become part of the standard set of techniques used by statisticians and other quantitative methodologists. The primary purpose of writing this monograph is to provide methodologists with techniques for conducting a Monte Carlo or simulation study. Although power method polynomials based on the standard normal distributions have been used in many contexts for the past 30 years, it was not until recently that the probability distribution and cumulative distribution functions were derived and made available to be used, for example, in the context of distribution fitting.

This monograph primarily concerns the use of power method polynomials in the context of simulating univariate and multivariate nonnormal distributions with specified cumulants and correlation matrices. The polynomials are easy to work with and will allow methodologists to simulate nonnormal distributions with cumulants and correlation structures in more elaborate situations, as demonstrated in Chapter 4.

This book is intended for statisticians, biostatisticians, and other quantitative methodologists in the social sciences, such as economics, education, psychology, etc. The monograph could also be used as a supplemental text for a graduate seminar course in Monte Carlo or simulation techniques, as there are many examples included that instructors and students can focus on and use as a guide. It is assumed that the reader has some knowledge of statistics, calculus, and linear algebra.

I thank Hakan Demirtas for his careful and thoughtful comments. I also acknowledge the help of Flaviu Hodis, who wrote some of the source code for the programs.

1

Introduction

The computer has made simulation and Monte Carlo methods widely applicable in statistical research. For example, in view of the plethora of evidence demonstrating that data can often be nonnormal (e.g., Blair, 1981; Bradley, 1968, 1982; Micceri, 1989; Pearson & Please, 1975; Sawilowsky & Blair, 1992), one of the primary purposes of using the method of Monte Carlo over the past few decades has been to investigate the properties of statistics such as the F-test in the context of the analysis of variance (ANOVA). Specifically, the typical statistics used to describe the (central) moments associated with a random variable are the mean, variance, skew, and kurtosis. The mean and variance describe the location and dispersion of the variable, and the values of skew and kurtosis describe its shape. In terms of the latter two statistics, Scheffe (1959) noted that the values of skew and kurtosis are "the most important indicators of the extent to which non-normality affects the usual inferences made in the analysis of variance" (p. 333). Indeed, the importance of this has been underscored by the voluminous amount of Monte Carlo investigations into the properties (e.g., Type I error and power) of the t or F statistics, where methodologists were required to simulate nonnormal distributions with specified values of skew and kurtosis.

Moreover, the need for empirical investigations associated with statistics has required statisticians to develop computationally efficient algorithms or transformations for generating pseudorandom numbers or for fitting distributions to data. For example, in terms of random variable generation, it may often be the case that a methodologist requires only an elementary transformation on standard normal or uniform deviates to create nonnormal data sets with specified values of skew and kurtosis. It is often the case that such transformations are based on a technique known as moment matching (see, for example, Devroye, 1986).

Fleishman (1978) introduced a widely used moment matching technique known as the power method for simulating univariate nonnormal distributions. This technique is based on a polynomial transformation that proceeds by taking the sum of a linear combination of a standard normal random variable, its square, and its cube. The power method allows for the systematic control of skew and kurtosis, which is needed in Monte Carlo studies, such as those in the context of ANOVA. This technique was extended from four to six moments (Headrick, 2002) and for simulating multivariate nonnormal distributions (Vale & Maurelli, 1983; Headrick & Sawilowsky, 1999a; Headrick, 2002). The power method has been recommended by other researchers. Specifically, in terms of simulating univariate nonnormal distributions, Tadikamalla (1980) suggested the power method could be used "if what is needed is a method to generate different distributions with the least amount of difficulty and as efficiently as possible" (p. 278). Similarly, in terms of multivariate data generation, Kotz, Balakrishnan, and Johnson (2000) noted: "This [power] method does provide a way of generating bivariate non-normal random variables. Simple extensions of other univariate methods are not available yet" (p. 37).

Power method polynomials have been used in studies from many different areas of research interest. Some examples in the context of the general linear model (e.g., ANCOVA, regression, repeated measures, and other univariate and multivariate (non)parametric tests) include studies by Beasley (2002), Beasley and Zumbo (2003), Finch (2005), Habib and Harwell (1989), Harwell and Serlin (1988, 1989), Headrick (1997), Headrick and Rotou (2001), Headrick and Sawilowsky (1999b, 2000a), Headrick and Vineyard (2001), Klockars and Moses (2002), Kowalchuk, Keselman, and Algina (2003), Lix, Algina, and Keselman (2003), Olejnik and Algina (1987), Rasch and Guiard (2004), Serlin and Harwell (2004), and Steyn (1993). Some examples of where power method transformations have been used in the context of measurement include computer adaptive testing (Zhu, Yu, & Liu, 2002), item response theory (Stone, 2003), and Markov Chain Monte Carlo (Bayesian) estimation (Hendrix & Habing, 2009). Other research topics or techniques where the power method has been used or where its properties have been studied include hierarchical linear models (Shieh, 2000), logistic regression (Hess, Olejnik, & Huberty, 2001), microarray analysis (Powell, Anderson, Cheng, & Alvord, 2002), multiple imputation (Demirtas & Hedeker, 2008), and structural equation modeling (Hipp & Bollen, 2003; Reinartz, Echambadi, & Chin, 2002; Welch & Kim, 2004).

Until recently, two problems associated with the power method were that its probability density function (pdf) and cumulative distribution function (cdf) were unknown (Tadikamalla, 1980; Kotz, Balakrishnan, & Johnson, 2000, p. 37). Thus, it was difficult (or impossible) to determine a power method distribution's percentiles, peakedness, tail-weight, or mode (Headrick & Sawilowsky, 2000b). However, these problems were resolved in an article by Headrick and Kowalchuck (2007), where the power method's pdf and cdf were derived in general form. Further, a power method system (Hodis & Headrick, 2007; Hodis, 2008) of distributions has been developed and is based on polynomial transformations that include the standard normal, logistic, and uniform distributions. Furthermore, recent developments have also demonstrated the power method to be useful for generating multivariate nonnormal distributions with specific types of structures. Some examples include continuous nonnormal distributions correlated with ranked or ordinal structures

(Headrick, 2004; Headrick, Aman, & Beasley, 2008a), systems of linear statistical equations (Headrick & Beasley, 2004), and distributions with specified intraclass correlations (Headrick & Zumbo, 2008).

There are other computationally efficient transformations that have demonstrated to be useful. Specifically, two other popular families based on moment matching techniques are the *g*-and-*h* family (Hoaglin, 1983, 1985; Headrick, Kowalchuk, & Sheng, 2008b; Martinez & Iglewicz, 1984; Morgenthaler & Tukey, 2000; Tukey, 1977) and the generalized lambda family of distributions (GLD) (Dudewicz & Karian, 1996; Freimer, Mudholkar, Kollia, & Lin, 1988; Headrick & Mugdadi, 2006; Karian & Dudewicz, 2000; Karian, Dudewicz, & McDonald, 1996; Ramberg & Schmeiser, 1972, 1974; Ramberg et al., 1979; Tukey, 1960).

The *g*-and-*h* family of distributions has been used in the context of statistical modeling of extreme events (Field & Genton, 2006) or in simulation studies that include such topics as common stock returns (Badrinath & Chatterjee, 1988, 1991), interest rate option pricing (Dutta & Babbel, 2005), portfolio management (Tang & Wu, 2006), stock market daily returns (Mills, 1995), extreme oceanic wind speeds (Dupuis & Field, 2004; Field & Genton, 2006), and regression, generalized additive models, or other applications of the general linear model (Keselman, Kowalchuk, & Lix, 1998; Keselman, Lix, & Kowalchuk, 1998; Keselman, Wilcox, Kowalchuk, & Olejnik, 2002; Keselman, Wilcox, Taylor, & Kowalchuk, 2000; Kowalchuk, Keselman, Wilcox, & Algina, 2006; Wilcox, 2001, 2006; Wilcox, Keselman, & Kowalchuk, 1998). Similarly, the GLD has received wide attention and has been used in studies that have included such topics or techniques as data mining (Dudewicz & Karian, 1999), independent component analysis (Karvanen, 2003; Mutihac & Van Hulle, 2003), microarray research (Beasley et al., 2004), operations research (Ganeshan, 2001), option pricing (Corrado, 2001), psychometrics (Bradley, 1993; Bradley & Fleisher, 1994; Delaney & Vargha, 2000), and structural equation modeling (Reinartz, Echambadi, & Chin, 2002).

It is perhaps not well-known that the power method, *g*-and-*h*, and GLD transformations are similar to the extent that they have general forms of a pdf and a cdf. More specifically, let $Q(V)$ denote the general form of the quantile function associated with the power method, *g*-and-*h*, or GLD transformation. The continuous variable V is standard normal in the context of both the power method and *g*-and-*h* transformations and is regular uniform in terms of the GLD transformation. The pdf and cdf associated with $Q(V)$ can be expressed in parametric form ($\mathbb{R}^2$) as

$$f_{Q(V)}(Q(v)) = f_{Q(V)}(Q(x,y)) = f_{Q(V)}\left(Q(v), \frac{f_V(v)}{Q'(v)}\right) \tag{1.1}$$

$$F_{Q(V)}(Q(v)) = F_{Q(V)}(Q(x,y)) = F_{Q(V)}(Q(v), F_V(v)) \tag{1.2}$$

where the derivative $Q'(v) > 0$, that is, $Q(V)$ is an increasing monotonic function. The specific forms of $f_V(v)$ and $F_V(v)$ are the standard normal or the regular uniform pdf and cdf, depending on the transformation being used.

To demonstrate the use of Equation 1.1, Figure 1.1 gives power method, g-and-h, and GLD approximations to a random sample of $N = 500$ observations drawn from a $t_{df=4}$ distribution with computed values of skew and kurtosis of –1.083 and 4.817, respectively.

The point being made here is that using the methodology and techniques developed in the subsequent chapters of this monograph allows researchers to evaluate the different transformations in terms of comparing percentiles, measures of central tendency (i.e., median, mode, trimmed means), goodness of fit tests, etc. Further, the transformations associated with each of the three specific forms of $Q(V)$ are simple to use because they only require algorithms for generating pseudorandom numbers (either standard normal or uniform) and a set of specified values that determine the shape of a nonnormal distribution that can be readily obtained from either tables (e.g., Karian & Dudewicz, 2000) or using available software that will numerically solve for these values (e.g., Headrick, Sheng, & Hodis, 2007; Headrick et al., 2008b; Kowalchuk & Headrick, 2009).

It is also important to point out that the power method, g-and-h, and GLD transformations are all capable of simulating multivariate nonnormal continuous distributions using a general approach (Headrick et al., 2008a; Kowalchuk & Headrick, 2008, 2009; Headrick & Mugdadi, 2006; Corrado, 2001). More importantly, this general procedure allows a user to generate correlated nonnormal distributions not only from within a family (e.g., the power method) but also between families. For example, a user may specify three nonnormal distributions with a specified correlation matrix where Distributions 1, 2, and 3 are based on power method, g-and-h, and GLD transformations, respectively. This general procedure has advantages because the monotonicity assumption in Equation 1.1, that is, that $Q'(v) > 0$, limits the parameter space of available values of skew and kurtosis for each of the three transformations. For example, the power method polynomials can generate distributions with moments (or cumulants) associated with the chi-square family of distributions ($df_{\geq 2}$) without violating the monotonicity criterion, whereas g-and-h transformations are unable to do so. On the other hand, g-and-h transformations can produce exact lognormal distributions, whereas the power method cannot. Thus, it is beneficial to have a procedure that is versatile enough that it allows a user to simulate different distributions from the three different families with controlled moments and correlations.

The primary focus of this monograph is on the theory and application of the power method transformation. The properties associated with the g-and-h and GLD families have been separately studied. Thus, the general relationship and characteristics that these two families have with the power method will be presented and discussed in the last chapter. What follows are brief descriptions of the content in each of the chapters in this monograph.

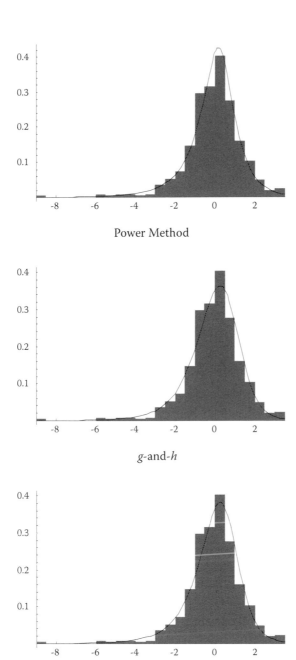

FIGURE 1.1
Power method, *g*-and-*h*, and GLD approximations of a random sample ($N = 500$) drawn from a $t_{df} = 4$ distribution with a skew of -1.083 and kurtosis of 4.817.

In Chapter 2 the general form of the power method's pdf and cdf for normal-, logistic-, and uniform-based polynomials is derived, as well as properties and indices associated with power method polynomials and pdfs such as measures of central tendency. Systems of equations that determine the coefficients for third- and fifth-order polynomials are developed, and the boundary conditions for determining if power method polynomials will yield valid pdfs are derived. *Mathematica*® functions available for implementing the various procedures, for example, solving coefficients, determining if a polynomial yields a valid pdf, calculating the height, mode, and trimmed mean of a pdf, graphing pdfs and cdfs, and so forth, are discussed. The limitation of available parameter space (i.e., combinations of skew and kurtosis) associated with power method transformations is discussed, and the general methodology for simulating multivariate correlated nonnormal distributions for the normal-, logistic-, and uniform-based power method polynomials is presented. Also, a step-by-step procedure to demonstrate the multivariate methodology using a numerical example is presented, with the steps for computing an intermediate correlation using *Mathematica* source code presented. Finally, closed-form equations for determining an intermediate correlation for normal-based polynomials are presented.

Chapter 3 demonstrates the methodology presented in Chapter 2. Using the *Mathematica* source code described in Section 2.4, examples of fitting power method pdfs to various theoretical pdfs such as the Beta, chi-square, Weibull, *F*, and *t* distributions are provided. These examples include superimposing power method pdfs on the theoretical pdfs, computing percentiles, and determining other various indices, such as modes, medians, and trimmed means. Third- and fifth-order polynomial approximations in terms of goodness of fit to the theoretical pdfs are compared and contrasted. Also, *Mathematica* source code is used to fit power method pdfs to empirical data sets. Numerous examples of logistic and uniform power method pdfs are provided, and remediation techniques associated with the limitation discussed in Section 2.5 are discussed. Chapter 3 also provides evidence from a Monte Carlo simulation to confirm that the methodology presented in Chapter 2 for the normal-, logistic-, and uniform-based polynomials generates nonnormal distributions with the specified cumulants and correlations. Finally, methods for improving the performance of a simulation based on power method polynomials are discussed.

In Chapter 4 a procedure for applying the power method in the context of simulating correlated nonnormal systems of linear statistical models based on fifth-order power method polynomials is developed. It is also demonstrated that the procedure allows for the simultaneous control of the correlated nonnormal (1) stochastic disturbances, (2) independent variables, and (3) dependent and independent variables for each model throughout a system. A procedure for simulating multivariate nonnormal distributions with specified intraclass correlation coefficients (ICCs) using the power method is developed, as well as a procedure for using the power method in the context of simulating controlled correlation structures between nonnormal (1) variates, (2) ranks, and (3) variates with ranks. The correlation structure between variates and their associated rank order is derived. Thus, a measure of the potential loss of information is provided when ranks are used in place of variates. Lastly, numerical examples and results from Monte Carlo simulations to demonstrate all of the procedures are provided.

In Chapter 5 the basic general univariate methodology for the g-and-h and GLD families of transformations is provided, as well as graphs of g-and-h and GLD pdfs, cdfs, and other numerical computations, such as percentiles, pdf heights, modes, trimmed means, and so forth, to demonstrate the univariate methodology for the transformations. Additionally, a method for approximating distributions for combined data sets when the underlying data sets are unavailable is provided. Finally, multivariate data generation for the g-and-h and GLD transformations is demonstrated, in addition to how power method polynomials, g-and-h, and GLD transformations can be used together for simulating correlated nonnormal variables.

2

The Power Method Transformation

2.1 Univariate Theory

The power method polynomial transformation produces a variety of distributions with specified standardized cumulants. For example, in the context of a third-order polynomial, the transformation proceeds by taking the sum of a linear combination of a standard normal, logistic, or uniform random variable, its square, and its cube (Hodis & Headrick, 2007; Hodis, 2008). The transformation is computationally efficient because it only requires the knowledge of a polynomial's coefficients and an algorithm that generates pseudorandom deviates.

We begin the derivation of the power method transformation's probability density function (pdf) and cumulative distribution function (cdf) with the following definitions.

Definition 2.1

Let W be a continuous random variable that follows either a standard (1) normal (Z), (2) logistic (L), or (3) uniform (U) distribution with pdf $f_W(w)$ defined as

$$f_W(w) = \begin{cases} f_{W:=Z}(w := z) = (2\pi)^{-\frac{1}{2}} \exp\{-z^2/2\}, & -\infty < z < +\infty \\ f_{W:=L}(w := l) = a\exp\{-al\}/(1+\exp\{-al\})^2, & -\infty < l < +\infty \\ f_{W:=U}(w := u) = 1/(2\sqrt{3}), & -\sqrt{3} < u < \sqrt{3} \end{cases} \quad (2.1)$$

where $a = \pi/\sqrt{3}$.

Definition 2.2

Let W have cdf $F_w(w)$ and moments μ_t generally defined as

$$F_W(w) = \Pr(W \le w) = \int_{-\infty}^{w} dF_W(w) \tag{2.2}$$

$$\mu_t = \mu_t(W) = \int_{-\infty}^{+\infty} w^t \, dF_W(w) \tag{2.3}$$

Definition 2.3

Let $w = (x, y)$ be the auxiliary variable that maps the parametric curves of the pdf and cdf associated with W as

$$f : z \to \mathbb{R}^2 := f_W(w) = f_W(x, y) = f_W(w, f_W(w)) \tag{2.4}$$

$$F : z \to \mathbb{R}^2 := F_W(w) = F_W(x, y) = F_W(w, F_W(w)) \tag{2.5}$$

Definition 2.4

Let Ω denote the family of sets of distributions with a finite number r of finite moments that define the power method transformation as $p : p(w) \to S_r$, where $\Omega = \{S_r : r \in \mathbb{N} = \{2, 4, 6, \dots, r\}$ and $S_2 \subseteq S_4 \subseteq S_6 \subseteq \cdots \subseteq S_r\}$. Let the analytical and empirical forms associated with p be expressed as

$$p(w) = \sum_{i=1}^{r} c_i w^{i-1} \tag{2.6}$$

$$p(W) = \sum_{i=1}^{r} c_i W^{i-1} \tag{2.7}$$

Definition 2.5

The transformation p is said to be a strictly increasing monotonic function in w with (1) $\deg(r - 1)$, (2) one and only one real root, (3) derivative $p'(w) > 0$, and (4) constant coefficients $c_i \in \mathbb{R}$, $c_r \neq 0$, and $\sum_{i=1}^{r/2} c_{2i} \mu_{2i} \in (0, 1]$ for all $p \in S_r$.

Definition 2.6

p is said to be a standard transformation where there exists a specified collection of r real-valued standardized moments θ and standardized cumulants γ such that $\gamma = f(\theta)$, and where the k-th moment is determined as

$$\theta_{1 \leq k \leq r} = E[p(W)^k] = E\left[\sum_{n=0}^{k} \binom{k}{n} c_1^{k-n} \left(\sum_{i=2}^{r} c_i W^{i-1}\right)^n\right] \tag{2.8}$$

for all $p \in S_r$.

The first and second moments are

$$\theta_1 = 0 = c_1 + \sum_{i=2}^{r} c_i \mu_{t=i-1} \tag{2.9}$$

$$\theta_2 = 1 = c_1^2 + \sum_{i=2}^{r} c_i^2 \mu_{t=2i-2} + 2 \sum_{i=1}^{r-1} \sum_{j=i+1}^{r} c_i c_j \mu_{t=i+j-2} \tag{2.10}$$

In more general notation, the k-th moment is expressed as

$$\theta_k = c_1^k + \sum_{i_1=2}^{r} c_{i_1}^k \mu_{t=ki_1-k} + \frac{k!}{n_1! n_2!} \sum_{i_1=1}^{r-1} \sum_{i_2=i_1+1}^{r} c_{i_1}^{n_1} c_{i_2}^{n_2} \mu_{t=n_1 i_1 + n_2 i_2 - k} + \cdots +$$

$$\frac{k!}{n_1! n_2! \ldots n_m!} \sum_{i_1=1}^{r-(m-1)} \sum_{i_2=i_1+1}^{r-(m-2)} \cdots \sum_{i_m=i_{(m-1)}+1}^{r} c_{i_1}^{n_1} c_{i_2}^{n_2} \ldots c_{i_m}^{n_m} \mu_{t=n_1 i_1 + n_2 i_2 + \cdots + n_m i_m - k}$$

$$+ \cdots + k! \sum_{i_1=1}^{r-(k-1)} \sum_{i_2=i_1+1}^{r-(k-2)} \cdots \sum_{i_k=i_{(k-1)}+1}^{r} c_{i_1} c_{i_2} \ldots c_{i_k} \mu_{t=i_1+i_2+\cdots+i_k-k} \tag{2.11}$$

where θ_k is composed of 2^{k-1} summands. We note that since W is associated with standard symmetric pdfs in (2.1), then all odd central moments are zero in (2.3), that is, $\mu_1 = \mu_3 = \cdots = \mu_{r(r-1)-1} = 0$, where $t = 1, \ldots, r(r-1)$.

Definition 2.7

The transformation p is said to have an associated system of r equations that express r specified standardized cumulants $\gamma_{1 \leq k \leq r}$ in terms of r coefficients, for all $p \in S_r$.

Proposition 2.1

If the compositions $f \circ p$ and $F \circ p$ map the parametric curves of $f_{p(W)}(p(w))$ and $F_{p(W)}(p(w))$ where $p(w) = p(x, y)$ as

$$f \circ p : p(w) \rightarrow \mathbb{R}^2 := f_{p(W)}(p(w)) = f_{p(W)}(p(x, y)) = f_{p(W)}(p(w), f_W(w)/p'(w)) \quad (2.12)$$

$$F \circ p : p(w) \rightarrow \mathbb{R}^2 := F_{p(W)}(p(w)) = F_{p(W)}(p(x, y)) = F_{p(W)}(p(w), F_W(w)) \quad (2.13)$$

then $f_{p(W)}(p(w), f_W(w)/p'(w))$ and $F_{p(W)}(p(w), F_W(w))$ in Equation 2.12 and Equation 2.13 are the pdf and cdf associated with the empirical form of the power method transformation $p(W)$ in Equation 2.7.

Proof

It is first shown that $f_{p(W)}(p(w), f_W(w)/p'(w))$ in Equation 2.12 has the following properties.

Property 2.1: $\displaystyle\int_{-\infty}^{+\infty} f_{p(W)}(p(w), f_W(w)/p'(w)) \, dw = 1$

Property 2.2: $f_{p(W)}(p(w), f_W(w)/p'(w)) \geq 0$

To prove Property 2.1, let $y = f(x)$ be a function where $\int_{-\infty}^{+\infty} f(x)dx = \int_{-\infty}^{+\infty} y \, dx$. Thus, given that $x = p(w)$ and $y = f_W(w)/p'(w)$ in $f_{p(W)}(p(x, y))$ in Equation 2.12, we have

$$\int_{-\infty}^{+\infty} f_{p(W)}(p(w), f_W(w)/p'(w)) \, dw = \int_{-\infty}^{+\infty} y \, dx = \int_{-\infty}^{+\infty} (f_W(w)/p'(w)) \, dp(w)$$

$$= \int_{-\infty}^{+\infty} (f_W(w)/p'(w)) \, p'(w) \, dw$$

$$= \int_{-\infty}^{+\infty} f_W(w) \, dw = 1$$

which integrates to 1 because $f_W(w)$ is either the standard normal, logistic, or uniform pdf. To prove Property 2.2, it is given by Definition 2.1 that $f_W(w) \geq 0$ and Definition 2.5 that $p'(w) > 0$. Hence, $f_{p(W)}(p(w), f_W(w)/p'(w)) \geq 0$ because $f_W(w)/p'(w)$ will be nonnegative in the space of w for all $w \in (-\infty, +\infty)$ and where $\lim_{w \to \pm\infty} f_W(w)/p'(w) = 0$ as $\lim_{w \to \pm\infty} f_W(w) = 0$ and $\lim_{w \to \pm\infty} 1/p'(w) = 0$.

A corollary to Proposition 2.1 is stated as follows:

Corollary 2.1

The derivative of the cdf $F_{p(W)}(p(w), F_W(w))$ in Equation 2.13 is the pdf $f_{p(W)}(p(w), f_W(w)/p'(w))$ in Equation 2.12.

Proof

It follows from $x = p(w)$ and $y = F_W(w)$ in $F_{p(W)}(p(x,y))$ in Equation 2.13 that $dx = p'(w)dw$ and $dy = f_W(w)dw$. Hence, using the parametric form of the derivative we have $y = dy/dx = f_W(w)/p'(w)$ in Equation 2.12. Thus, $F'_{p(W)}(p(w), F_W(w)) = F'_{p(W)}(p(x, dy/dx)) = f_{p(W)}(p(x, y)) = f_{p(W)}(p(w), f_W(w)/p'(w))$. Therefore, $f_{p(W)}(p(w), f_W(w)/p'(w))$ in Equation 2.12 and $F_{p(W)}(p(w), F_W(w))$ in Equation 2.13 are the pdf and cdf associated with $p(W)$ in Equation 2.7.

Some other properties associated with the power method are subsequently provided.

Property 2.3

The median of the power method's pdf in Equation 2.12 is located at $p(w) = c_1$. This can be shown by letting $x_{0.50} = p(w)$ and $y = 0.50 = F_W(w) = \Pr\{W \le w\}$ denote the coordinates of the cdf in Equation 2.13 that are associated with the 50th percentile. In general, we must have $w = 0$ such that $y = 0.50 = F_W(0) = \Pr\{W \le 0\}$ holds in Equation 2.13 for the standard normal, logistic, and uniform pdfs in Equation 2.1. Thus, taking the limit of the polynomial as $w \to 0$ locates the median at $\lim_{w \to 0} p(w) = c_1$.

Property 2.4

The 100α percent symmetric trimmed mean of the power method's pdf in Equation 2.12 is located at $p(w) = (1 - 2\alpha)^{-1} \sum_{i=1}^{r} c_i \int_{F_W^{-1}(\alpha)}^{F_W^{-1}(1-\alpha)} w^{i-1} dF_W(w)$, where $\alpha \in (0, 0.50)$. This follows from the usual definition for a symmetric trimmed mean, for example, $T(F) = \mu_{trim} = (1 - 2\alpha)^{-1} \int_{F^{-1}(\alpha)}^{F^{-1}(1-\alpha)} x dF(x)$ (Bickel & Doksum, 1977, p. 401) and that the power method's first moment in Equation 2.9 can alternatively be expressed as $\theta_1 = 0 = \sum_{i=1}^{r} c_i \int_{-\infty}^{+\infty} w^{i-1} dF_W(w)$. As such, as $\alpha \to 0$ (or $\alpha \to 0.5$), the trimmed mean will approach the mean $p(w) \approx 0$ (or median $p(w) \approx c_1$).

Property 2.5

A mode associated with Equation 2.12 is located at $f_{p(W)}(p(\tilde{w}), f_W(\tilde{w})/p'(\tilde{w}))$, where $w = \tilde{w}$ is the critical number that solves $dy/dw = d(f_W(w)/p'(w))/dw = 0$ and either globally or locally maximizes $y = f_W(\tilde{w})/p'(\tilde{w})$ at $x = p(\tilde{w})$. Further, if $W := Z$ or $W := L$ in Equation 2.1, then the power method pdf will have a global maximum. This is due to the fact that the standard normal and logistic pdfs have global maximums and the transformation $p(w)$ is a strictly increasing monotonic function.

Property 2.6

An index of strength of the positive relationship between W and $p(W)$ in Equation 2.7 is the product-moment coefficient of correlation (ρ), where $\rho_{W,p(W)} = \sum_{i=1}^{r/2} c_{2i}\mu_{2i}$ and $\rho_{W,p(W)} \in (0,1]$, as specified in Definition 2.5. This can be shown by setting $\rho_{W,p(W)} = E[Wp(W)]$ since W and $p(W)$ have zero means and unit variances. Thus, we have $E[Wp(W)] = \sum_{i=1}^{r} c_i E[W^i] = \sum_{i=1}^{r/2} c_i c_{2i}\mu_{2i}$ because all odd central moments associated with W are zero. We note that the special case of where $\rho_{W,p(W)} = 1$ occurs when $r = 2$ and where $c_1 = 0$ and $c_2 = 1$, that is, the pdfs in Equation 2.1.

Property 2.7

If all odd subscripted coefficients in Equation 2.7 are zero, that is, $c_{2i-1} = 0$ for all $i = 1,\ldots,r/2$, then the odd cumulants will be zero and thus the power method distribution is symmetric. Further, simultaneous sign reversals between nonzero values of these coefficients (c_{2i-1}) will reverse the signs of the odd cumulants but will have no effect on the even cumulants. For example, if $r = 4$, simultaneously reversing the signs between c_1 and c_3 will change the direction of skew (e.g., from positive to negative) but will have no effect on the variance or kurtosis. This can be most readily seen by inspecting the systems of equations for $r = 4$ and $r = 6$ that we are concerned with in subsequent sections.

In summary, this section provided a general derivation of the power method's pdf and cdf. As such, a user is able to calculate probabilities and other measures of central tendency that are associated with valid power method pdfs. Some other useful properties were also provided, such as, for ($r = 4$), a user need only consider solving for the coefficients in Equation 2.7 for distributions with positive skew. A distribution with negative skew can be easily obtained by simultaneously reversing the signs of the odd subscripted coefficients. In the next two sections we will consider third-order ($r = 4$) and fifth-order ($r = 6$) systems in more detail.

2.2 Third-Order Systems

The first set of distributions to be considered is S_4, that is, Equation 2.7, with $r = 4$ because the set S_2 consists of only a linear transformation on the random variable W in Equation 2.1. A third-order system of equations, based on Definition 2.7, allows a user to control the first four cumulants of a power method distribution. To derive the system of equations for determining the coefficients associated with the set S_4 requires obtaining the standardized moments $\mu_{1 \leq k \leq 12}$ for Equation 2.1 using Equation 2.3 and subsequently substituting these moments into the expansions of $p(W)^{1 \leq k \leq 4}$, which yields the moments $\theta_{1 \leq k \leq 4} = E[p(W)^{1 \leq k \leq 4}]$ associated with Equation 2.8. As such, the equations for the mean (γ_1), variance (γ_2), skew (γ_3), and kurtosis (γ_4) are in general

$$\gamma_1 = \theta_1 = 0 = c_1 + c_3 \tag{2.14}$$

$$\gamma_2 = \theta_2 = 1 = c_2^2 + (\mu_4 - 1)c_3^2 + \mu_4 2c_2c_4 + \mu_6 c_4^2 \tag{2.15}$$

$$\gamma_3 = \theta_3 = (\mu_4 - 1)3c_2^2 c_3 + (\mu_6 - 3\mu_4 + 2)c_3^3 + (\mu_6 - \mu_4)6c_2c_3c_4 + (\mu_8 - \mu_6)3c_3c_4^2 \tag{2.16}$$

$$\gamma_4 = \theta_4 - 3 = \mu_{12}c_4^4 + \mu_4 c_2^4 + \mu_6 4c_2^3 c_4 + (\mu_{10} - 2\mu_8 + \mu_6)6c_3^2 c_4^2$$
$$+ (\mu_8 - 4\mu_6 + 6\mu_4 - 3)c_3^4 + 6c_2^2((\mu_6 - 2\mu_4 + 1)c_3^2 + \mu_8 c_4^2) \tag{2.17}$$
$$+ 4c_2c_4(\mu_{10}c_4^2 + (\mu_8 - 2\mu_6 + \mu_4)3c_3^2) - 3$$

The specific forms of Equation 2.14 through Equation 2.17 and the relevant even moments associated with the standard normal (Z), logistic (L), and uniform (U) pdfs are

$$\gamma_{1(Z)} = 0 = c_1 + c_3 \tag{2.18}$$

$$\gamma_{2(Z)} = 1 = c_2^2 + 2c_3^2 + 6c_2c_4 + 15c_4^2 \tag{2.19}$$

$$\gamma_{3(Z)} = 8c_3^3 + 6c_2^2 c_3 + 72c_2c_3c_4 + 270c_3c_4^2 \tag{2.20}$$

$$\gamma_{4(Z)} = 3c_2^4 + 60c_2^2 c_3^2 + 60c_3^4 + 60c_2^3 c_4 + 936c_2c_3^2 c_4 + 630c_2^2 c_4^2 + 4500c_3^2 c_4^2$$
$$+ 3780c_2c_4^3 + 10395c_4^4 - 3 \tag{2.21}$$

where $\mu_4 = 3$, $\mu_6 = 15$, $\mu_8 = 105$, $\mu_{10} = 945$, and $\mu_{12} = 10395$;

$$\gamma_{1(L)} = 0 = c_1 + c_3 \tag{2.22}$$

$$\gamma_{2(L)} = 1 = c_2^2 + 16c_3^2/5 + 42c_2c_4/5 + 279c_4^2/7 \tag{2.23}$$

$$\gamma_{3(L)} = 48c_2^2c_3/5 + 1024c_3^3/35 + 7488c_2c_3c_4/35 + 67824c_3c_4^2/35 \qquad (2.24)$$

$$\gamma_{4(L)} = 21c_2^4/5 + 6816c_2^2c_3^2/35 + 3840c_3^4/7 + 1116c_2^3c_4/7 + 51264c_2c_3^2c_4/7$$

$$+ 20574c_2^2c_4^2/5 + 40384224c_3^2c_4^2/385 + 827820c_2c_4^3/11 + 343717911c_4^4/455 - 3$$
$$(2.25)$$

where $\mu_4 = 21/5$, $\mu_6 = 279/7$, $\mu_8 = 3429/5$, $\mu_{10} = 206955/11$, and $\mu_{12} = 343717911/455$; and

$$\gamma_{1(U)} = 0 = c_1 + c_3 \qquad (2.26)$$

$$\gamma_{2(U)} = 1 = c_2^2 + 4c_3^2/5 + 18c_2c_4/5 + 27c_4^2/7 \qquad (2.27)$$

$$\gamma_{3(U)} = 4\left(75075c_2^2c_3 + 14300c_3^3 + 386100c_2c_3c_4 + 482625c_3c_4^2\right)/125125 \qquad (2.28)$$

$$\gamma_{4(U)} = 9c_2^4/5 + 48c_3^4/35 + 108c_2^3c_4/7 + 3672c_2^2c_4^2/77 + 729c_4^4/13 + 264c_2^2c_3^2/35$$
$$+ 972c_2c_4^3/11 + 1296c_2c_3^2c_4/35 + 54c_2^2c_4^2 - 3 \qquad (2.29)$$

where $\mu_4 = 9/5$, $\mu_6 = 27/7$, $\mu_8 = 9$, $\mu_{10} = 243/11$, and $\mu_{12} = 729/13$.

In summary, the three systems of equations above consist of four equations expressed in terms of four variables $(c_{i=1,\ldots,4})$. The first two equations in each system are set such that any power method distribution will have a mean of zero and unit variance. The next two equations in each system are set to a user's specified values of skew and kurtosis. Simultaneously solving a system yields the solution values of the coefficients for a third-order polynomial.

Figure 2.1 gives the graphs of some examples of third-order power method pdfs and cdfs based on the standard (1) normal, (2) logistic, and (3) uniform systems of equations above. The associated standardized cumulants and coefficients for these graphs are provided in Table 2.1. The coefficients were determined by solving the third-order systems of equations using the source code discussed in Section 2.4.

From Definition 2.5, a polynomial must have a derivative such that $p'(w) > 0$. Thus, it is subsequently shown, in the context of third-order polynomials, what combinations of skew and kurtosis will yield valid pdfs for each of the distributions in Equation 2.1. We begin the derivation by solving $p'(w) = 0$, which gives

$$w = \left(-c_3 \pm \left(c_3^2 - 3c_2c_4\right)^{\frac{1}{2}}\right)\Big/(3c_4) \qquad (2.30)$$

Solving Equation 2.15 for c_3^2 and substituting this expression into the discriminant of Equation 2.30 and equating this result to zero yields the binary quadratic form

$$1 = c_2^2 + (5\mu_4 - 3)c_2c_4 + \mu_6c_4^2 \qquad (2.31)$$

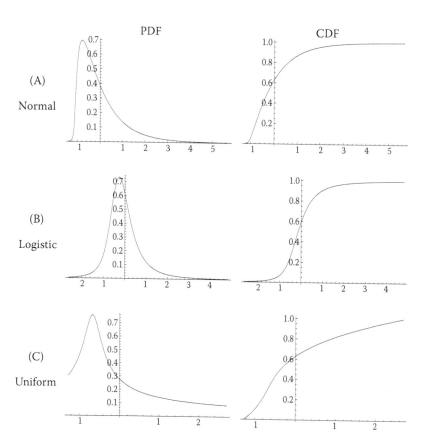

FIGURE 2.1
Examples of third-order power method distributions. The pdfs and cdfs are based on polynomial transformations from the standard (A) normal, (B) logistic, and (C) uniform distributions. See Table 2.1 for the cumulants and coefficients.

Equation 2.31 gives the points where the discriminant vanishes in Equation 2.30, and thus where real values of w exist such that $p'(w) = 0$. Solving Equation 2.31 for c_4 yields the principal root

$$c_4 = \left(3c_2 - \mu_4 5c_2 + ((3c_2 - \mu_4 5c_2)^2 + \mu_6 4(1 - c_2)^2)^{\frac{1}{2}}\right)\Big/(2\mu_6) \qquad (2.32)$$

TABLE 2.1

Cumulants (γ_i) and Coefficients (c_i) for the Third-Order PDFs in Figure 2.1

(A) Normal		(B) Logistic		(C) Uniform	
$\gamma_1 = 0$	$c_1 = -0.280908$	$\gamma_1 = 0$	$c_1 = -0.138784$	$\gamma_1 = 0$	$c_1 = -0.356908$
$\gamma_2 = 1$	$c_2 = 0.791924$	$\gamma_2 = 1$	$c_2 = 0.648952$	$\gamma_2 = 1$	$c_2 = 0.625202$
$\gamma_3 = 2$	$c_3 = 0.280908$	$\gamma_3 = 3$	$c_3 = 0.138784$	$\gamma_3 = 1$	$c_3 = 0.356908$
$\gamma_4 = 6.5$	$c_4 = 0.040164$	$\gamma_4 = 50$	$c_4 = 0.064482$	$\gamma_4 = 0$	$c_4 = 0.173669$

Solving Equation 2.15 for c_3 and substituting this expression and Equation 2.32 into Equation 2.16 and Equation 2.17 yields the general boundary conditions for skew $\bar{\gamma}_3$ and kurtosis $\bar{\gamma}_4$ as

$$\bar{\gamma}_3 = \left(2\mu_6^2\right)^{-1}(3(3/2)^{\frac{1}{2}}((c_2(3c_2 - \mu_4 5c_2 + ((3c_2 - \mu_4 5c_2)^2 - \mu_6 4(c_2^2 - 1))^{\frac{1}{2}})/\mu_6)^{\frac{1}{2}}$$

$$(2\mu_6(\mu_8 - \mu_6) + c_2^2((9 - 13\mu_4)\mu_6^2 + \mu_6(5\mu_4 - 2\mu_8 - 3) + \mu_8(3 - 5\mu_4)) \qquad (2.33)$$

$$- c_2((25\mu_4^2 - 4\mu_6 - 30\mu_4 + 9)c_2^2 + 4\mu_6)^{\frac{1}{2}}(\mu_8(5\mu_4 - 3) + \mu_6 - 3\mu_6^2)))$$

$$\bar{\gamma}_4 = (1/(\mu_4 - 1)^2)(12\mu_4 - 3\mu_4^2 - 4\mu_6 + \mu_8 - 6 + c_2^4(3 + 10\mu_4^2 + \mu_4^3 + 2\mu_6$$

$$-\mu_4(6\mu_6 + 11) + \mu_8) + (1/\mu_6)(2c_2^3(3c_2 - 5c_2\mu_4 + ((3c_2 - 5c_2\mu_4)^2$$

$$-4(c_2^2 - 1)\mu_6)^{\frac{1}{2}})(6\mu_4^3 - 5\mu_6 - 2\mu_4^2(\mu_6 + 3) + \mu_4(3 + 3\mu_6 - 2\mu_8) + 3\mu_8))$$

$$+(1/2\mu_6^2)(3c_2 - 5c_2\mu_4 + ((3c_2 - 5c_2\mu_4)^2 - 4(c_2^2 - 1)\mu_6)^{\frac{1}{2}})^2(3\mu_{10}(\mu_4 - 1)$$

$$+4\mu_6^2 + 6\mu_8 - \mu_6\mu_8 - 3\mu_4(\mu_6 + 2\mu_8))) + (1/16\mu_6^4)((3c_2 - 5c_2\mu_4$$

$$+((3c_2 - 5c_2\mu_4)^2 - 4(c_2^2 - 1)\mu_6)^{\frac{1}{2}})^4(\mu_{12}(\mu_4 - 1)^2 + \mu_6(12(\mu_4 - 1)\mu_8$$

$$-6\mu_{10}(\mu_4 - 1) - 4\mu_6^2 + \mu_6(\mu_8 + 3)))) - 2c_2^2(6\mu_4^2 - 3\mu_4 - \mu_6 - 3\mu_4\mu_6 + \mu_8$$

$$+(1/4\mu_6^2)((3c_2 - 5c_2\mu_4 + ((3c_2 - 5c_2\mu_4)^2 - 4(c_2^2 - 1)\mu_6)^{\frac{1}{2}})^2(3\mu_{10}(\mu_4 - 1)$$

$$+3\mu_4(10\mu_6 + \mu_6^2 - 4\mu_8) + (\mu_6 - 3)(\mu_6 - \mu_8) + \mu_4^2(7\mu_8 - 6 - 22\mu_6))))$$

$$-(1/\mu_6)(2c_2(3c_2 - 5c_2\mu_4 + ((3c_2 - 5c_2\mu_4)^2 - 4(c_2^2 - 1)\mu_6)^{\frac{1}{2}})$$

$$\times(3\mu_4^2 - 6\mu_6 + 2\mu_4(\mu_6 - \mu_8) + 3\mu_8 - (1/4\mu_6^2)((3c_2 - 5c_2\mu_4 + ((3c_2 - 5c_2\mu_4)^2$$

$$-4(c_2^2 - 1)\mu_6)^{\frac{1}{2}})^2(\mu_{10}(1 + \mu_4 - 2\mu_4^2) + 6\mu_4^2\mu_8 + 3\mu_6(\mu_8 - 2\mu_6)$$

$$+\mu_4(3\mu_6 + 2\mu_6^2 - 6\mu_8 - 2\mu_6\mu_8))))))$$

$$(2.34)$$

For the pdfs considered in Equation 2.1, Equation 2.33 is equal to zero and satisfies the constraints that $\gamma_1 = 0$ and $\gamma_2 = 1$, when either $c_2 = 0$ or $c_2 = 1$. Further, nonzero real values of $\bar{\gamma}_3$ exist only when $0 < c_2 < 1$. Thus, a third-order power method pdf is valid when we satisfy the conditions that (1) c_4 is strictly greater than the right-hand side of Equation 2.32, which implies that the discriminant in Equation 2.30 is less than zero, and (2) $0 < c_2 < 1$. We note that if $c_2 = 1$, then the power method pdfs reduce to the set S_2 and are the standard pdfs in 2.1 because the distributions are symmetric (i.e., $c_1 = c_3 = 0$) and c_4 must equal zero for Equation 2.31 to hold.

Figure 2.2 gives the graphs of the regions for valid third-order power method pdfs for the distributions considered in Equation 2.1. These graphs were obtained using the specific forms of Equation 2.33 and Equation 2.34 for the standard normal (Z), logistic (L), and uniform (U) distributions, which are

$$\overline{\gamma}_{3(Z)} = 2\left(\frac{3}{5}c_2\left((15+21c_2^2)^{\frac{1}{2}}(27+332c_2^2+508c_2^4)-6c_2(87+392c_2^2+388c_2^4)\right)\right)^{\frac{1}{2}} \quad (2.35)$$

$$\overline{\gamma}_{4(Z)} = \frac{8}{125}\left(675-5040c_2^2-13101c_2^4+c_2(15+21c_2^2)^{1/2}(2866c_2^2+45)\right) \quad (2.36)$$

$$\overline{\gamma}_{3(L)} = \frac{16}{4805}\left(\frac{1}{217}c_2\left((217+224c_2^2)^{\frac{1}{2}}\right.\right.$$

$$\times\left(1492324407+18710779120c_2^2+25333499648c_2^4\right)$$

$$\left.\left.-21c_2\left(5403659755+22077243836c_2^2+18055699584c_2^4\right)\right)\right)^{\frac{1}{2}} \quad (2.37)$$

$$\overline{\gamma}_{4(L)} = \frac{6}{420202055}\left(33093052699+65202201792c_2^2-45746265600c_2^4\right.$$

$$\left.+c_2(217+224c_2^2)^{\frac{1}{2}}(3025815552c_2^2-5524146432)\right) \quad (2.38)$$

$$\overline{\gamma}_{3(U)} = \frac{4}{35\sqrt{3}}\left(c_2(21+28c_2^2)^{\frac{1}{2}}(1225+10920c_2^2+1376c_2^4)\right.$$

$$\left.-c_2^2(23275+85260c_2^2+72912c_2^4)\right)^{\frac{1}{2}} \quad (2.39)$$

$$\overline{\gamma}_{4(U)} = \frac{2}{1365}\left(525-21280c_2^2-38976c_2^4+32c_2(21+28c_2^2)^{1/2}(228c_2^2+35)\right) \quad (2.40)$$

The regions inside the enclosed boundaries depicted in Figure 2.2 represent the asymmetric distributions that are elements in the set S_4, that is, valid pdfs. Table 2.2 gives the minimum and maximum values of γ_3 and γ_4 for the three power method transformations.

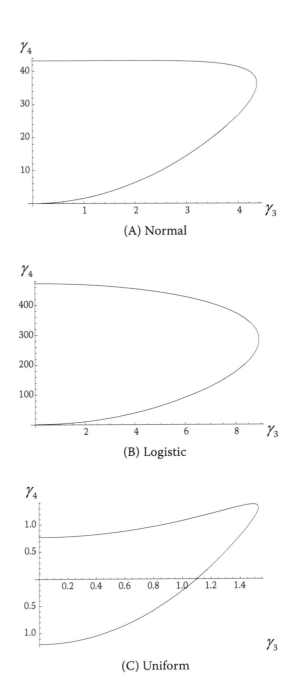

(A) Normal

(B) Logistic

(C) Uniform

FIGURE 2.2
Parameter space for valid third-order pdfs in the skew (γ_3) and kurtosis (γ_4) plane. For asymmetric distributions, valid pdfs exist in the regions inside the graphed boundaries. See Table 2.2 for minimum and maximum values of γ_3 and γ_4 for each transformation. Points on the boundaries represent coordinates of γ_3 and γ_4 associated with Equations 2.35–2.40.

TABLE 2.2

Boundary Values of Skew $|\bar{\gamma}_3|$ and Kurtosis $\bar{\gamma}_4$ for Third-Order Power Method PDFs

Distribution	c_2	c_4	$|\bar{\gamma}_3|$	$|\bar{\gamma}_3|$
Normal	1	0	0	0^a
	0	0.258199	0	43.2
	0.018387	0.250905	2.051991	43.300410^b
	0.213476	0.180918	4.363294^b	36.343090
Logistic	1	0	0	1.2^a
	0	0.158397	0	472.530569^b
	0.191132	0.118197	8.913658^b	283.962261
Uniform	1	0	0	-1.2^a
	0	0.509175	0	0.769231
	0.197279	0.368780	1.490033	1.376586^b
	0.283894	0.315029	1.528637^b	1.311615

[a] Indicates a lower limit.

[b] Indicates an upper limit that was obtained by numerically solving for the optimal value of c_2 that maximizes Equation 2.35 through Equation 2.40. The value of c_4 is determined by evaluating a specific form of Equation 2.32 for the optimal value of c_2.

2.3 Fifth-Order Systems

One of the implications of the requirement that $p'(w) > 0$ in Definition 2.5 is that it restricts the parameter space in terms of the various combinations of skew and kurtosis for valid power method pdfs as evidenced by the regions graphed in Figure 2.2. For example, consider the third-order system of equations based on the standard normal distribution. This system will not yield valid power method pdfs for values of skew and kurtosis associated with the important chi-square family of distributions or any other combination of skew and kurtosis where $\gamma_3^2/\gamma_4 > 9/14$ (Headrick & Kowalchuk, 2007, Property 4.4). Thus, it becomes necessary to choose the largest set of S_r available, subject to the constraints that it remains computationally efficient to (1) solve the system of r equations for the coefficients in Equation 2.7 and (2) determine whether or not a power method distribution has a valid pdf.

In view of the above, the largest set to be considered herein is S_6 because it is well known that solutions to $p'(w) = 0$ for $r > 6$ cannot be expressed in terms of radicals, and thus, closed-formed solutions are not available. As such, the first four cumulants associated with the set S_6 are

$$\gamma_1 = \theta_1 = 0 = c_1 + c_3 + \mu_4 c_5 \tag{2.41}$$

$$\gamma_2 = \theta_2 = 1 = c_2^2 + (\mu_4 - 1)c_3^2 + \mu_6 c_4^2 + 2(\mu_6 - \mu_4)c_3 c_5 - \mu_4^2 c_5^2 + \mu_8 c_5^2 + 2\mu_8 c_4 c_6 + \mu_{10} c_6^2$$
$$+ 2c_2(\mu_4 c_4 + \mu_6 c_6)$$

$$\tag{2.42}$$

$$\gamma_3 = \theta_3 = (2 - 3\mu_4 + \mu_6)c_3^3 + 3(-(\mu_4 - 2)\mu_4 - 2\mu_6 + \mu_8)c_3^2 c_5$$
$$+ 3c_2^2((\mu_4 - 1)c_3 + (\mu_6 - \mu_4)c_5) + 3c_3\left((\mu_{10} + 2\mu_4^2)c_5^2 - \mu_6(c_4^2 + 2\mu_4 c_5^2)\right)$$
$$+ c_6(2\mu_{10}c_4 - \mu_{10}c_6 + \mu_{12}c_6) + \mu_8\left(c_4^2 - c_6^2 - 2c_4 c_6\right))$$
$$+ 6c_2\left(\mu_8 c_4 c_5 - \mu_4^2 c_4 c_5 + \mu_6 c_3(c_4 - c_6) + \mu_8 c_3 c_6 + \mu_{10}c_5 c_6\right.$$
$$- \mu_4(c_3 c_4 + \mu_6 c_5 c_6)) + c_5\left(\mu_{12}c_5^2 + 2\mu_4^3 c_5^2 + 6\mu_{12}c_4 c_6 + 3\mu_{14}c_6^2\right.$$
$$+ 3\mu_{10}\left(c_4^2 - \mu_4 c_6^2\right) - 3\mu_4\left(\mu_6 c_4^2 + \mu_8(c_5^2 + 2c_4 c_6)\right))$$

(2.43)

$$\gamma_4 = \theta_4 - 3 = (\mu_8 - 4\mu_6 - 3)c_3^4 + \mu_{12}c_4^4 + 4(\mu_{10} + 3\mu_6 - 3\mu_8)c_3^3 c_5 + 6\mu_{14}c_4^2 c_5^2$$
$$+ \mu_{16}c_5^4 - 3\mu_4^4 c_5^4 + 6\mu_4^3 c_5^2\left(c_3^2 + 2c_2 c_4 - 2c_3 c_5\right) + 4\mu_{14}c_4^3 c_6 + 12\mu_{16}c_4 c_5^2 c_6$$
$$+ 6\mu_{16}c_4^2 c_6^2 + 6\mu_{18}c_5^2 c_6^2 + 4\mu_{18}c_4 c_6^3 + \mu_{20}c_6^4 + 4c_3^3(\mu_6 c_4 + \mu_8 c_6) + 6c_2^2\left((\mu_6 + 1)c_3^2\right.$$
$$+ \mu_8 c_4^2 + 2(\mu_8 - \mu_6)c_3 c_5 + \mu_{10}c_5^2 + 2\mu_{10}c_4 c_6 + \mu_{12}c_6^2) + 6c_3^2\left(\mu_{10}c_4^2 + \mu_6 c_4^2 - 2\mu_8 c_4^2\right.$$
$$- 2\mu_{10}c_5^2 + \mu_{12}c_5^2 + \mu_8 c_5^2 + 2(\mu_{12} + \mu_8 - 2\mu_{10})c_4 c_6 + (\mu_{10} - 2\mu_{12} + \mu_{14})c_6^2)$$
$$+ 4c_3 c_5\left(3\mu_{12}c_4^2 - 3\mu_{10}c_4^2 - \mu_{12}c_5^2 + 6(\mu_{14} - \mu_{12})c_4 c_6 + 3(\mu_{16} - \mu_{14})c_6^2\right)$$
$$+ 4c_2\left(c_4\left(\mu_{10}c_4^2 - 6\mu_6 c_3^2 + 3\mu_8 c_3(c_3 - 2c_5^2) + 6\mu_{10}c_3 c_5 + 3\mu_{12}c_5^2\right)\right.$$
$$+ 3((\mu_{10} + \mu_6 - 2\mu_8)c_3^2 + \mu_{12}c_4^2 + 2(\mu_{12} - \mu_{10})c_3 c_5 + \mu_{14}c_5^2)c_6$$
$$+ 3\mu_{14}c_4 c_6^2 + \mu_{16}c_6^3) + 6\mu_4^2 c_5\left(2c_3^3 - 3c_3^2 c_5 + 2\mu_6 c_3 c_5^2 + c_2^2(c_5 - 2c_3)\right.$$
$$+ 2c_2(2c_3 c_4 + \mu_6 c_5 c_6) + c_5\left(\mu_6 c_4^2 + \mu_{10}c_6^2 + \mu_8(c_5^2 + 2c_4 c_6)\right))$$
$$+ \mu_4\left(c_2^4 - 12c_2^2\left(c_3^2 - c_3 c_5 + \mu_6 c_5^2\right) + 12c_2\left(c_3^2 c_4 - 2c_5^2(\mu_8 c_4 + \mu_{10}c_6)\right.\right.$$
$$- 2c_3 c_5(\mu_6(c_4 - c_6) + \mu_8 c_6)) - 2((6 + 2\mu_6)c_3^3 c_5 - 3c_3^4 + 6(\mu_8 - 2\mu_6)c_3^3 c_5^2$$
$$+ 6c_3 c_5\left(\mu_{10}c_5^2 - \mu_6 c_4^2 + c_6(2\mu_{10}c_4 - \mu_{10}c_6 + \mu_{12}c_6) + \mu_8\left(c_4^2 - c_6^2 - 2c_4 c_6\right)\right)$$
$$+ 2c_5^2\left(3\mu_{10}c_4^2 + 3\mu_{14}c_6^2 + \mu_{12}\left(c_5^2 + 6c_4 c_6\right)\right)))) - 3$$

(2.44)

Using γ_3 and γ_4, the fifth and six standardized cumulants are generally expressed as

$$\gamma_5 = \theta_5 - 10\gamma_3 \tag{2.45}$$

$$\gamma_6 = \theta_6 - 15\gamma_4 - 10\gamma_3^2 - 15 \tag{2.46}$$

The specific forms of the expressions for γ_5 and γ_6 are large. See, for example, Headrick (2002, Equations B3 and B4), Headrick and Kowalchuk (2007, Equations A5 and A6), or Headrick and Zumbo (2008, Equations 12 and 13) in the context of standard normal-based polynomials.

In order to verify that a set of solved coefficients $(c_1, \ldots, c_6)$ for a fifth-order polynomial yields a valid pdf, we will need to evaluate the closed form formulae that solve the equation $p'(w) = 0$. More specifically, if the coefficients satisfy Property 2.6 and the complex numbers of w associated with the following formulae

$$w = \pm \left(\left(\pm \left(9c_4^2 - 20c_2c_6 \right)^{\frac{1}{2}} - 3c_4 \right) \Big/ 10c_6 \right)^{\frac{1}{2}} \quad \text{and}$$

$$w = \mp \left(\left(\pm \left(9c_4^2 - 20c_2c_6 \right)^{\frac{1}{2}} - 3c_4 \right) \Big/ 10c_6 \right)^{\frac{1}{2}} \tag{2.47}$$

$$w = \pm \frac{1}{2} D^{\frac{1}{2}} \pm \frac{1}{2} (E \pm F)^{\frac{1}{2}} - \frac{c_5}{5c_6} \quad \text{and} \quad w = \pm \frac{1}{2} D^{\frac{1}{2}} \mp \frac{1}{2} (E \pm F)^{\frac{1}{2}} - \frac{c_5}{5c_6} \tag{2.48}$$

for symmetric distributions Equation 2.47 or for asymmetric distributions Equation 2.48 have nonzero imaginary parts, then $p'(w) > 0$ and thus $p(w) \in S_6$ will have a valid pdf. The expressions for D, E, and F in Equation 2.48 are as follows:

$$D = 4c_5^2/25c_6^2 + \left(B^{\frac{1}{2}} + A \right)^{\frac{1}{3}} \Big/ \left(15c_6 2^{\frac{1}{3}} \right) + 2^{\frac{1}{3}} C \Big/ \left(5c_6 \left(B^{\frac{1}{2}} + A \right)^{\frac{1}{3}} \right) - 2c_4/5c_6$$

$$E = 8c_5^2/25c_6^2 - \left(B^{\frac{1}{2}} + A \right)^{\frac{1}{3}} \Big/ \left(15c_6 2^{\frac{1}{3}} \right) - 2^{\frac{1}{3}} C \Big/ \left(5c_6 \left(B^{\frac{1}{2}} + A \right)^{\frac{1}{3}} \right) 4c_4 / 5c_6$$

$$F = D^{-\frac{1}{2}} \left(48c_4c_5/100c_6^2 - 16c_3/20c_6 - 64c_5^3/500c_6^3 \right)$$

where

$$A = 54c_4^3 - 216c_3c_4c_5 + 432c_2c_5^2 + 540c_3^2c_6 - 1080c_2c_4c_6$$
$$B = A^2 - 4(9c_4^2 - 24c_3c_5 + 60c_2c_6)^3$$
$$C = 3c_4^2 - 8c_3c_5 + 20c_2c_6$$

Provided in Figure 2.3 and Table 2.3 are examples of graphs, cumulants, and coefficients for some fifth-order power method pdfs and cdfs. These graphs and solutions were obtained using the source code discussed in the next section. Inspection of Figure 2.3 reveals that fifth-order polynomials can have some advantages over third-order polynomials. Specifically, fifth-order polynomials based on the unit normal pdf have the ability to generate some distributions with negative kurtosis (Figure 2.3A). See Section 3.2 for more examples of comparisons between third- and fifth-order polynomials. Further, polynomials based on the standard uniform distribution can generate distributions that may have more than one mode (Figure 2.3C), whereas third-order polynomials based on the uniform distribution will have only one mode, which is located at $p(w) = -c_3/(3c_4)$ from Property 2.5.

Another advantage that fifth-order systems have, when juxtaposed to third-order systems, is that a larger region in the skew and kurtosis plane (e.g., Figure 2.2A) is captured. For example, contrary to the third-order normal-based system, the fifth-order system will provide valid power method pdfs for the cumulants associated with the chi-square family of distributions

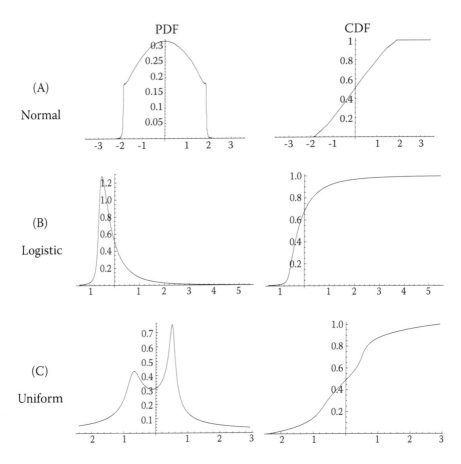

FIGURE 2.3
Examples of fifth-order power method distributions. The pdfs and cdfs are based on polyno-mials transformations from the standard (A) normal, (B) logistic, and (C) uniform distribu-tions. See Table 2.3 for the associated cumulants and coefficients.

TABLE 2.3

Cumulants (γ_i) and Coefficients (c_i) for the Fifth-Order PDFs in Figure 2.3

(A) Normal		(B) Logistic		(C) Uniform	
$\gamma_1 = 0$	$c_1 = 0$	$\gamma_1 = 0$	$c_1 = -0.258947$	$\gamma_1 = 0$	$c_1 = 0.037636$
$\gamma_2 = 1$	$c_2 = 1.248344$	$\gamma_2 = 1$	$c_2 = 0.517374$	$\gamma_2 = 1$	$c_2 = 0.914329$
$\gamma_3 = 0$	$c_3 = 0$	$\gamma_3 = 6$	$c_3 = 0.264295$	$\gamma_3 = 1/4$	$c_3 = -0.209385$
$\gamma_4 = -1$	$c_4 = -0.111426$	$\gamma_4 = 200$	$c_4 = 0.063461$	$\gamma_4 = 1/4$	$c_4 = -0.419859$
$\gamma_5 = 0$	$c_5 = 0$	$\gamma_5 = 7500$	$c_5 = -0.001273$	$\gamma_5 = 0$	$c_5 = 0.095416$
$\gamma_6 = 48/7$	$c_6 = 0.004833$	$\gamma_6 = 7500000$	$c_6 = 0.000677$	$\gamma_6 = -7/2$	$c_6 = 0.209131$

with $df \geq 2$. More generally, this system will provide valid power method pdfs for the small family of distributions with cumulants of $\gamma_4 = (3/2)\gamma_3^2$, $\gamma_5 = 3\gamma_3^3$, and $\gamma_6 = (15/2)\gamma_3^4$, where $0 \leq \gamma_3 \leq 2$ and subsumes the cumulants for chi-square distributions with $df \geq 2$. As such, if $\gamma_3 = 2$ (or $\gamma_3 = 0$), then the power method pdf will have the cumulants associated with a chi-square distribution with $df = 2$ (or the unit normal distribution).

2.4 *Mathematica*® Functions

To implement the methodology developed in the previous sections, *Mathematica* (Wolfram, 2003, version 5.2) functions, source code, and demonstrations are available at the *Journal of Statistical Software* Web site, http://www. jstatsoft.org/v19/i03, for fifth-order normal-based polynomials. Specifically, listed in Table 2.4 are brief descriptions of the functions associated with the

TABLE 2.4

Mathematica Functions for the Power Method Transformation

Function	Usage
TheoreticalCumulants	Computes standardized cumulants for a pdf
EmpiricalCumulantsM(K)	Computes standardized cumulants for a set of data based on method of moments (or Fisher's k-statistics)
PowerMethod1	Computes polynomial coefficients based on an asymmetric pdf or a set of data
PowerMethod2	Computes polynomial coefficients based on a symmetric pdf
PowerMethod3	Computes polynomial coefficients based on a set of standardized cumulants specified by the user
Pcorr	Computes the Pearson correlation between w and $p(w)$
AsymmetricWs	Computes the solutions to $p'(w) = 0$ for asymmetric power method polynomials
SymmetricWs	Computes the solutions to $p'(w) = 0$ for symmetric power method polynomials
CheckPdf	Determines if a set of coefficients yield a valid or invalid power method pdf
PlotProbDist	Plots the graph of a valid or invalid power method pdf
HeightPdf	Computes the maximum height of a valid power method pdf
ModePdf	Computes the mode of a valid power method pdf
PlotCumulativeDist	Plots the cdf for a valid or invalid power method pdf
SuperImposeTheoretical	Superimposes a power method pdf on a theoretical pdf
SuperImposeEmpirical	Superimposes a power method pdf on an empirical pdf
CumulativeProb	Computes a cumulative probability for a power method pdf
TrimMean	Computes a symmetric trimmed mean for a power method pdf
InterCorr	Computes an intermediate correlation between two specified normal-based power method distributions

Mathematica software package developed by Headrick, Sheng, and Hodis (2007). This software package has also been modified to include fifth-order polynomials based on either the uniform or logistic distributions as well as third-order polynomials with any of the underlying distributions introduced in Equation 2.1. These software packages are available at http://www.siu.edu/~epse/headrick/PowerMethod/.

2.5 Limitations

As mentioned in the Section 2.3, one of the limitations associated with power method polynomials is that not all combinations of skew (γ_3) and kurtosis (γ_4) are possible. More generally, this limitation can be summarized by the classic moment problem expressed as follows (Devroye, 1986, p. 682):

$$\begin{vmatrix} 1 & \mu_1 & \mu_2 & \cdots & \mu_i \\ \mu_1 & \mu_2 & & & \mu_{i+1} \\ \mu_2 & & & & \mu_{i+2} \\ \vdots & & & & \vdots \\ \mu_i & \mu_{i+1} & \mu_{i+2} & \cdots & \mu_{2i} \end{vmatrix} \geq 0 \qquad (2.49)$$

Without loss of generality, if we consider the first four standardized moments (i.e., setting $\mu_1 = 0$ and $\mu_2 = 1$), the inequality in Equation 2.49 implies that $\mu_4 \geq \mu_3^2 + 1$ (or $\gamma_4 \geq \gamma_3^2 - 2$), where equality is associated with a Bernoulli variable. Inspection of Figure 2.2 indicates that not all combinations of skew and kurtosis are possible for third-order polynomials. For example, given a value of skew $\gamma_3 = 2$, a distribution must have a value of kurtosis that is greater than 2. However, a normal-based third-order power method polynomial requires a value of kurtosis that is greater than (approximately) 6.34 to produce a valid pdf.

In the context of symmetric distributions, extending Equation 2.49 to six moments implies that $\mu_6 \geq \mu_4^2$ (or $\gamma_6 \geq \gamma_4^2 - 9\gamma_4 - 6$). As such, consider a symmetric fifth-order power method distribution with $\mu_4 = 4$ ($\gamma_4 = 1$) based on the standard normal distribution. To produce a valid power method pdf for this value of kurtosis would require the sixth moment (cumulant) to be slightly greater than $\mu_6 = 36$ ($\gamma_6 = 6$), which is greater than the lower-boundary value of $\mu_6 = 16$ ($\gamma_6 = -14$). This limitation is also present in the context of asymmetric power method distributions. Some techniques that can be used to ameliorate these limitations are discussed in the Section 3.3.

Another limitation associated with fifth-order polynomials is that, unlike skew (γ_3) and kurtosis (γ_4), the descriptive statistics associated with the fifth and six cumulants (γ_5 and γ_6) are not computed by most modern software packages. Thus, if a theoretical density (e.g., a standard exponential pdf where $\gamma_5 = 24$ and $\gamma_6 = 120$) is not being simulated, it may be difficult to select initial values of γ_5 and γ_6 for Equation 2.45 and Equation 2.46. An approach a user could take to overcome this limitation is to use values of γ_5 and γ_6 associated with a third-order system as initial starting values. Specifically, solve a third-order system for specified values of γ_3 and γ_4 and subsequently evaluate Equation 2.45 and Equation 2.46 using the solutions of $c_1, ..., c_4$, and set $c_5 = c_6 = 0$, which will yield the values of γ_5 and γ_6 associated with the third-order polynomial. For example, listed in Table 2.3 are the values of $\gamma_3 = 1/4$ and $\gamma_4 = 1/4$ for the distribution based on the uniform pdf. Using the coefficients for the third-order polynomial and evaluating Equation 2.45 and Equation 2.46 with $c_5 = c_6 = 0$ yields the starting values of $\gamma_5 = -0.818$ and $\gamma_6 = -5.672$, which can then be manipulated to create the bimodal distribution depicted in Figure 2.3C.

2.6 Multivariate Theory

Let $W_1, ..., W_T$ be continuous random variables as given in Definition 2.1. If W_j and W_k are defined to be standard normal, that is, $W_j := Z_j$ and $W_k := Z_k$ in Equation 2.1, then their distribution functions and bivariate density function can be expressed as

$$\Phi(z_j) = \Pr\{Z_j \le z_j\} = \int_{-\infty}^{z_j} (2\pi)^{-\frac{1}{2}} \exp\left\{-u_j^2/2\right\} du_j \tag{2.50}$$

$$\Phi(z_k) = \Pr\{Z_k \le z_k\} = \int_{-\infty}^{z_k} (2\pi)^{-\frac{1}{2}} \exp\left\{-u_k^2/2\right\} du_k \tag{2.51}$$

$$f_{jk} := f_{Z_j Z_k}(z_j, z_k, \rho_{Z_j Z_k}) = \left(2\pi\left(1 - \rho_{Z_j Z_k}^2\right)^{\frac{1}{2}}\right)^{-1}$$
$$\exp\left\{-\left(2\left(1 - \rho_{Z_j Z_k}^2\right)\right)^{-1}\left(z_j^2 - 2\rho_{Z_j Z_k} z_j z_k + z_k^2\right)\right\} \tag{2.52}$$

Using Equations 2.50 and 2.51, it follows that the transformations

$$W_j := L_j = (\sqrt{3}/\pi)\ln(\Phi(z_j)/(1 - \Phi(z_j))) \tag{2.53}$$

$$W_k := U_k = \sqrt{3}(2\Phi(z_k) - 1) \tag{2.54}$$

where $\Phi(z_j)$ and $\Phi(z_k) \sim U[0,1]$, will have logistic and uniform distributions with the associated standard pdfs defined in Equation 2.1, respectively. As such, if the polynomials $p(W_j)$ and $p(W_k)$ are based on either the normal, logistic, or the uniform distributions in Equation 2.1, then their bivariate correlation can be determined as

$$\rho_{p(W_j),p(W_k)} = \int_{-\infty}^{+\infty} \int_{-\infty}^{+\infty} \left(\sum_{i=1}^{r} c_{ji} w_j^{i-1} \right) \left(\sum_{i=1}^{r} c_{ki} w_k^{i-1} \right) f_{jk} \, dz_j \, dz_k \qquad (2.55)$$

where the correlation $\rho_{z_j z_k}$ in f_{jk} is referred to as an intermediate correlation.

Given the above, the following are steps a user may take to simulate multivariate distributions with specified standardized cumulants and intercorrelations:

1. Specify the standardized cumulants for T power method polynomials $p(W_1),...,p(W_T)$ and obtain the coefficients for each polynomial. Specify a $T \times T$ correlation matrix for the power method distributions.

2. Compute the intermediate correlations by using the coefficients computed in Step 1 and Equation 2.55. For two variables, this is accomplished by iteratively altering the intermediate correlation $\rho_{z_j z_k}$ in (2.55) until $\rho_{p(W_j),p(W_k)}$ has the specified correlation given in Step 1. Repeat this step for all $T(T-1)/2$ pairwise combinations. Note that if the polynomials are based on the standard logistic or uniform distributions, then Equation 2.53 or 2.54 will also be needed in the computation. See Table 2.5 for an example using *Mathematica* (Wolfram, 2003) source code.

3. Assemble the intermediate correlations determined in Step 2 into a $T \times T$ matrix and decompose this matrix using a Cholesky factorization. Note that this requires the intermediate correlation matrix to be sufficiently positive definite.

4. Use the results from Step 3 to generate T standard normal variables $(Z_1,...,Z_T)$ correlated at the intermediate levels determined in Step 2 as

$$Z_1 = a_{11} V_1$$
$$Z_2 = a_{12} V_1 + a_{22} V_2$$
$$\vdots$$
$$Z_j = a_{1j} V_1 + a_{2j} V_2 + \cdots + a_{ij} V_i + \cdots + a_{jj} V_j \qquad (2.56)$$
$$\vdots$$
$$Z_T = a_{1T} V_1 + a_{2T} V_2 + \cdots + a_{iT} V_i + \cdots + a_{jT} V_j + \cdots + a_{TT} V_T$$

where $V_1,...,V_T$ are independent standard normal random deviates and where a_{ij} represents the element in the i-th row and j-th column of the matrix associated with the Cholesky factorization.

5. Using $Z_1,....,Z_T$ from Equation 2.56, generate the T power method distributions specified in Step 1. If any of the polynomials are based on the logistic or uniform distributions, then it is suggested that the required uniform deviates in either Equation 2.53 or 2.54 be generated using

TABLE 2.5

Mathematica Source Code for Determining an Intermediate Correlation

(* Intermediate correlation *)
$\rho_{Z_1 Z_2} = 0.674825$;
(* Standard normal cdf—Equation 2.51 *)
$\Phi = \int_{-\infty}^{z_2} \text{Sqrt}[2*\pi]^{-1} * \text{Exp}[-u^2/2]du$;
(* Standard logistic quantile function—Equation 2.53 *)
$l = \text{Sqrt}[3/\pi^2] * \text{Log}(\Phi/(1-\Phi))$;
(* Polynomial based on the standard normal distribution—Equation 2.7 *)
$Y_1 = c_{11} + c_{12}z_1 + c_{13}z_1^2 + c_{14}z_1^3 + c_{15}z_1^4 + c_{16}z_1^5$;
(* Polynomial based on the standard logistic distribution—Equation 2.7 *)
$Y_2 = c_{21} + c_{22}l + c_{23}l^2 + c_{24}l^3 + c_{25}l^4 + c_{26}l^5$;
(* Standard bivariate normal pdf—Equation 2.52 *)
$f_{12} = (2*\pi*(1-\rho_{Z_1 Z_2}^2)^{\frac{1}{2}})^{-1} * \text{Exp}[-(2*(1-\rho_{Z_1 Z_2}^2))^{-1} * (z_1^2 - 2*\rho_{Z_1 Z_2} * z_1 * z_2 + z_2^2)]$;
(* Correlation between Y_1 and Y_2—Equation 2.55 *)
rho = NIntegrate[$Y_1 * Y_2 * f_{12}$, {z_1, –8, 8}, {z_2, –8, 8}, Method $\to$ "Method"]
(* The intermediate correlation $\rho_{Z_1 Z_2} = 0.674825$ yields the specified correlation *rho* *)
rho = 0.50

See panels A and B in Table 2.3 for values of the coefficients for the polynomials.

the following series expansion for the unit normal cdf (Marsaglia, 2004):

$$\Phi(z) = (1/2) + \varphi(z)\{z + z^3/3 + z^5/(3 \cdot 5) + z^7/(3 \cdot 5 \cdot 7) + z^9/(3 \cdot 5 \cdot 7 \cdot 9) + \cdots\} \quad (2.57)$$

where $\varphi(z)$ is the standard normal pdf and where the absolute error associated with Equation 2.57 is less than 8×10^{-16}.

To demonstrate the computations in Steps 2 and 3, suppose we desire that the three distributions depicted in Figure 2.3 have the following intercorrelations: $\rho_{p(Z),p(L)} = 0.50$, $\rho_{p(Z),p(U)} = 0.70$, and $\rho_{p(L),p(U)} = 0.60$. Following Step 2, and using the coefficients listed in Table 2.3, the intermediate correlations are $\rho_{Z_1 Z_2} = 0.674825$, $\rho_{Z_1 Z_3} = 0.715175$, and $\rho_{Z_2 Z_3} = 0.760712$, where the subscripts denote that the polynomials are based on the $1 = Z$, $2 = L$, and $3 = U$ distributions. Table 2.5 gives an example for the computation of $\rho_{Z_1 Z_2}$. The results of a Cholesky factorization on the intermediate correlation matrix yield $a_{11} = 1$, $a_{12} = 0.674825$, $a_{22} = 0.737978$, $a_{13} = 0.715175$, $a_{23} = 0.376833$, and $a_{33} = 0.588661$ for (2.56).

It is worthy to point out that for the special case of when $W_j := Z_j$ and $W_k := Z_k$, for all $W_1, \ldots, W_T$, there is an alternative to using Equation 2.55 for computing intermediate correlations. More specifically, we can derive an equation for determining intermediate correlations by making use of Equation 2.56 for $T = 2$. That is, $Z_2 = a_{12}Z_1 + a_{22}V_2$, where $a_{12} \in (-1, 1)$, and $Z_1 = V_1$, since $a_{11} = 1$, and thus we have the correlation $\rho_{Z_1 Z_2} = a_{12}$. As such, because $p(Z_1)$ and $p(Z_2)$ have zero means and unit variances, their correlation can be determined as

$$\rho_{p(Z_1), p(Z_2)} = E[p(Z_1)p(Z_2)] = E[p(Z_1)p(a_{12}Z_1 + a_{22}V_2)] \tag{2.58}$$

Expanding the right-hand side of Equation 2.58 and taking expectations by substituting the moments from the standard normal distribution, we have, in more general notation, for fifth-order polynomials (Headrick, 2002)

$$\begin{aligned}
\rho_{p(Z_j), p(Z_k)} &= 3c_{5j}c_{1k} + 3c_{5j}c_{3k} + 9c_{5j}c_{5k} + c_{1j}(c_{1k} + c_{3k} + 3c_{5k}) + c_{2j}c_{2k}\rho_{Z_j Z_k} \\
&\quad + 3c_{4j}c_{2k}\rho_{Z_j Z_k} + 15c_{6j}c_{2k}\rho_{Z_j Z_k} + 3c_{2j}c_{4k}\rho_{Z_j Z_k} + 9c_{4j}c_{4k}\rho_{Z_j Z_k} \\
&\quad + 45c_{6j}c_{4k}\rho_{Z_j Z_k} + 15c_{2j}c_{6k}\rho_{Z_j Z_k} + 45c_{4j}c_{6k}\rho_{Z_j Z_k} + 225c_{6j}c_{6k}\rho_{Z_j Z_k} \\
&\quad + 12c_{5j}c_{3k}\rho_{Z_j Z_k}^2 + 72c_{5j}c_{5k}\rho_{Z_j Z_k}^2 + 6c_{4j}c_{4k}\rho_{Z_j Z_k}^3 + 60c_{6j}c_{4k}\rho_{Z_j Z_k}^3 \\
&\quad + 60c_{4j}c_{6k}\rho_{Z_j Z_k}^3 + 600c_{6j}c_{6k}\rho_{Z_j Z_k}^3 + 24c_{5j}c_{5k}\rho_{Z_j Z_k}^4 + 120c_{6j}c_{6k}\rho_{Z_j Z_k}^5 \\
&\quad + c_{3j}\left(c_{1k} + c_{3k} + 3c_{5k} + 2c_{3k}\rho_{Z_j Z_k}^2 + 12c_{5k}\rho_{Z_j Z_k}^2\right)
\end{aligned}$$

$$\tag{2.59}$$

where (2.59) can be solved using the *Mathematica* function InterCorr listed in Table 2.4. Setting $c_5 = c_6 = 0$ in Equation 2.59 and simplifying gives (Vale & Maurelli, 1983; Headrick & Sawilowsky, 1999a)

$$\begin{aligned}
\rho_{p(Z_j), p(Z_k)} &= \rho_{Z_j Z_k} \\
&\left(c_{2j}c_{2k} + 3c_{4j}c_{2k} + 3c_{2j}c_{4k} + 9c_{4j}c_{4k} + 2c_{1j}c_{1k}\rho_{Z_j Z_k} + 6c_{4j}c_{4k}\rho_{Z_j Z_k}^2\right)
\end{aligned} \tag{2.60}$$

for computing intermediate correlations associated with third-order polynomials. As in Step 2, if there are T polynomials, then Equation 2.59 or Equation 2.60 can be used to solve for all $T(T-1)/2$ intermediate correlations.

3

Using the Power Method Transformation

3.1 Introduction

The primary focus of this chapter is to demonstrate the use of the source code discussed in Section 2.4 and in Headrick, Sheng, and Hodis (2007) by making use of the *Mathematica* functions listed in Table 2.4 in the context of both univariate and multivariate settings. Specifically, graphical illustrations of third- and fifth-order normal-, logistic-, and uniform-based power method pdfs are provided as well as examples of computing percentiles and other indices, such as measures of central tendency. Some comparisons are also made between power method pdfs and theoretical (empirical) distributions. Empirical estimates of power method distributions cumulants and correlation coefficients are provided to demonstrate and confirm the methodology presented in Chapter 2.

Another focus of this chapter is to consider remediation techniques to address the limitation associated with power method polynomials that was discussed in Section 2.5: The system of power method distributions does not span the entire space in the skew (γ_3) and kurtosis (γ_4) plane, that is, $\gamma_4 \geq \gamma_3^2 - 2$. For example, the fifth and sixth standardized cumulants associated with any chi-square distribution $(df = k)$ can be expressed as a function of γ_3 as $\gamma_5 = 3\gamma_3^3$ and $\gamma_6 = (15/2)\gamma_3^4$, where $\gamma_3 = (8/k)^{\frac{1}{2}}$. Given these cumulants, fifth-order power method polynomials cannot produce valid pdfs with values of kurtosis such that $\gamma_4 < 12/k$. More specifically, if we were to set $k = 2$, then we would have $\gamma_3 = 2$, $\gamma_5 = 24$, and $\gamma_6 = 120$, and thus, given these values, we could not obtain valid pdfs for values of kurtosis that are less than 6. As such, some techniques are presented and discussed to assist the user to ameliorate this limitation. Other techniques associated with improving the speed of a simulation and the efficiency of estimates are also discussed.

3.2 Examples of Third- and Fifth-Order Polynomials

The moments or cumulants associated with theoretical distributions (e.g., beta, chi-square, Weibull, etc.) are often used in Monte Carlo or simulation studies. Let us first consider a real-valued continuous stochastic variable X with distribution function F. The central moments of X can be defined as

$$\mu_r = \mu_r(X) = \int_{-\infty}^{+\infty} (x - \mu)^r dF(x) \tag{3.1}$$

Further, if σ is defined as the population standard deviation associated with X, then using Equation 3.1, the first $r = 6$ standardized cumulants are $\gamma_1 = 0, \gamma_2 = 1,$

$$\gamma_3 = \mu_3/\sigma^3 \tag{3.2}$$

$$\gamma_4 = \mu_4/\sigma^4 - 3 \tag{3.3}$$

$$\gamma_5 = \mu_5/\sigma^5 - 10\gamma_3 \tag{3.4}$$

$$\gamma_6 = \mu_6/\sigma^6 - 15\gamma_4 - 10\gamma_3^2 - 15 \tag{3.5}$$

Tables 3.1–3.15 provide some examples of commonly used theoretical distributions. Included are (superimposed) approximations of these theoretical pdfs made by third- and fifth-order normal-based power method polynomials with their corresponding cumulants, coefficients, percentiles, and other indices. Inspection of these tables indicates that fifth-order polynomials will often provide more accurate approximations to theoretical pdfs than the third-order polynomials, for example, the beta, χ_3^2, and Weibull (2,5) distributions.

One of the factors inhibiting the precision of third-order polynomials is that the values of kurtosis (γ_4) had to be increased to ensure valid power method pdfs on several occasions. For example, consider the beta and F distributions in Tables 3.1–3.3 and Tables 3.10–3.12, respectively. Inspection of the pdfs and percentiles in these tables indicates that the third-order polynomials more accurately approximate the F distributions than the beta distributions. In terms of the beta distributions, all values of kurtosis had to be substantially increased to ensure valid third-order power method pdfs, whereas only one value of kurtosis had to be slightly increased in the context of the F distribution approximations ($F_{3,100}$). A criterion for evaluating an exogenous increase in kurtosis is discussed in Section 3.3.

It is also worth noting that the fifth and sixth cumulants (γ_5, γ_6) associated with third-order polynomials can differ markedly from the theoretical pdfs or the fifth-order polynomial cumulants. The fifth and sixth cumulants for a third-order polynomial can be determined by substituting its solved coefficients ($c_1, \ldots, c_4; c_5 = c_6 = 0$) into the equations for γ_5 and γ_6 associated with the

TABLE 3.1

Third- and Fifth-Order Standard Normal Power Method PDF Approximations (Dashed Lines) to a Beta(2,4) Distribution

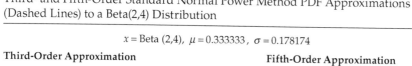

$$x = \text{Beta } (2,4), \; \mu = 0.333333, \; \sigma = 0.178174$$

Third-Order Approximation

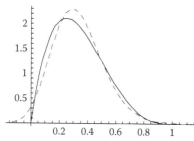

Fifth-Order Approximation

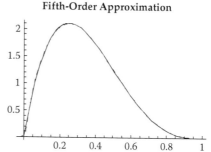

Standardized Cumulants (γ_i)			Coefficients (c_i)	
x	Third	Fifth	Third	Fifth
$\gamma_1 = 0$	$\gamma_1 = 0$	$\gamma_1 = 0$	$c_1 = -0.077243$	$c_1 = -0.108304$
$\gamma_2 = 1$	$\gamma_2 = 1$	$\gamma_2 = 1$	$c_2 = 0.987418$	$c_2 = 1.104253$
$\gamma_3 = 0.467707$	$\gamma_3 = 0.467707$	$\gamma_3 = 0.467707$	$c_3 = 0.077243$	$c_3 = 0.123347$
$\gamma_4 = -0.375$	$\hat{\gamma}_4 = 0.345$	$\gamma_4 = -0.375$	$c_4 = 0.002195$	$c_4 = -0.045284$
$\gamma_5 = -1.403122$	$\hat{\gamma}_5 = 0.352302$	$\gamma_5 = -1.403122$	$c_5 = 0.0$	$c_5 = -0.005014$
$\gamma_6 = -0.426136$	$\hat{\gamma}_6 = 0.462921$	$\gamma_6 = -0.426136$	$c_6 = 0.0$	$c_6 = 0.001285$

pdf	x	Third	Fifth
Height	2.1094	2.2958	2.1044
Mode	0.25	0.2921	0.2507
Median	0.3138	0.3197	0.3140
20% trimmed mean	0.3186	0.3225	0.3187
Percentiles			
0.01	0.0327	−0.1095	0.0351
0.05	0.0764	0.0656	0.0765
0.25	0.1938	0.2075	0.1936
0.75	0.4542	0.4443	0.4541
0.95	0.6574	0.6477	0.6574
0.99	0.7779	0.8090	0.7785

TABLE 3.2

Third- and Fifth-Order Standard Normal Power Method PDF Approximations
(Dashed Lines) to a Beta(4,4) Distribution

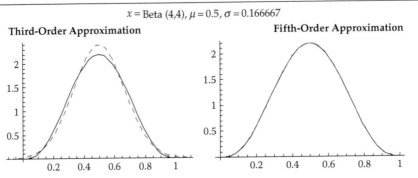

$$x = \text{Beta (4,4)}, \mu = 0.5, \sigma = 0.166667$$

	Standardized Cumulants (γ_i)		Coefficients (c_i)	
x	Third	Fifth	Third	Fifth
$\gamma_1 = 0$	$\gamma_1 = 0$	$\gamma_1 = 0$	$c_1 = 0.0$	$c_1 = 0.0$
$\gamma_2 = 1$	$\gamma_2 = 1$	$\gamma_2 = 1$	$c_2 = 0.999433$	$c_2 = 1.093437$
$\gamma_3 = 0.0$	$\gamma_3 = 0.0$	$\gamma_3 = 0.0$	$c_3 = 0.0$	$c_3 = 0.0$
$\gamma_4 = -0.545455$	$\hat{\gamma}_4 = 0.004545$	$\gamma_4 = -0.545455$	$c_4 = 0.000189$	$c_4 = -0.035711$
$\gamma_5 = 0.0$	$\hat{\gamma}_5 = 0.0$	$\gamma_5 = 0.0$	$c_5 = 0.0$	$c_5 = 0.0$
$\gamma_6 = 1.678322$	$\hat{\gamma}_6 = 0.000116$	$\gamma_6 = 1.678322$	$c_6 = 0.0$	$c_6 = 0.000752$

pdf	x	Third	Fifth
Height	2.1875	2.2958	2.1891
Mode	0.5	0.5	0.5
Median	0.5	0.5	0.5
20% trimmed mean	0.5	0.5	0.5
Percentiles			
0.01	0.1423	0.1105	0.1424
0.05	0.2253	0.2260	0.2252
0.25	0.3788	0.3881	0.3789
0.75	0.6212	0.6119	0.6211
0.95	0.7747	0.7740	0.7748
0.99	0.8577	0.8895	0.8576

TABLE 3.3

Third- and Fifth-Order Standard Normal Power Method PDF Approximations (Dashed Lines) to a Beta(5,4) Distribution

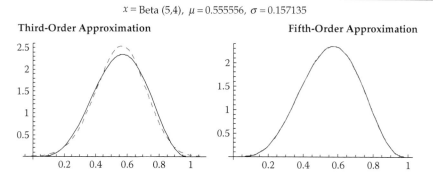

x = Beta (5,4), $\mu = 0.555556$, $\sigma = 0.157135$

	Standardized Cumulants (γ_i)		Coefficients (c_i)	
x	Third	Fifth	Third	Fifth
$\gamma_1 = 0$	$\gamma_1 = 0$	$\gamma_1 = 0$	$c_1 = 0.021377$	$c_1 = 0.027142$
$\gamma_2 = 1$	$\gamma_2 = 1$	$\gamma_2 = 1$	$c_2 = 0.998210$	$c_2 = 1.081901$
$\gamma_3 = -0.128565$	$\gamma_3 = -0.128565$	$\gamma_3 = -0.128565$	$c_3 = -.021377$	$c_3 = -0.029698$
$\gamma_4 = -0.477273$	$\hat{\gamma}_4 = 0.032727$	$\gamma_4 = -0.477273$	$c_4 = 0.000444$	$c_4 = -0.0311452$
$\gamma_5 = 0.341191$	$\hat{\gamma}_5 = -0.011595$	$\gamma_5 = 0.341191$	$c_5 = 0.0$	$c_5 = 0.000852$
$\gamma_6 = 1.242053$	$\hat{\gamma}_6 = 0.005393$	$\gamma_6 = 1.242053$	$c_6 = 0.0$	$c_6 = 0.000600$

pdf	x	Third	Fifth
Height	2.3500	2.5457	2.3509
Mode	0.5714	0.5656	0.5711
Median	0.5598	0.5593	0.5598
20% trimmed mean	0.5588	0.5582	0.5588
Percentiles			
0.01	0.1982	0.1314	0.1984
0.05	0.2892	0.2613	0.2892
0.25	0.4445	0.4397	0.4446
0.75	0.6709	0.6755	0.6709
0.95	0.8071	0.7774	0.8072
0.99	0.8791	0.8371	0.8788

TABLE 3.4

Third- and Fifth-Order Standard Normal Power Method PDF Approximations
(Dashed Lines) to χ_3^2 Distribution

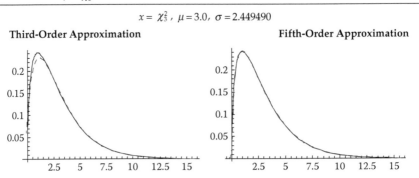

$$x = \chi_3^2, \ \mu = 3.0, \ \sigma = 2.449490$$

Third-Order Approximation   Fifth-Order Approximation

	Standardized Cumulants (γ_i)		Coefficients (c_i)	
x	Third	Fifth	Third	Fifth
$\gamma_1 = 0$	$\gamma_1 = 0$	$\gamma_1 = 0$	$c_1 = -0.247155$	$c_1 = -0.259037$
$\gamma_2 = 1$	$\gamma_2 = 1$	$\gamma_2 = 1$	$c_2 = 0.862899$	$c_2 = 0.867102$
$\gamma_3 = 1.632993$	$\gamma_3 = 1.632993$	$\gamma_3 = 1.632993$	$c_3 = 0.247155$	$c_3 = 0.265362$
$\gamma_4 = 4.0$	$\hat{\gamma}_4 = 4.21$	$\gamma_4 = 4.0$	$c_4 = 0.024057$	$c_4 = 0.021276$
$\gamma_5 = 13.063945$	$\hat{\gamma}_5 = 15.110100$	$\gamma_5 = 13.063945$	$c_5 = 0.0$	$c_5 = -0.002108$
$\gamma_6 = 53.333333$	$\hat{\gamma}_6 = 70.216428$	$\gamma_6 = 53.333333$	$c_6 = 0.0$	$c_6 = 0.000092$

pdf	x	Third	Fifth
Height	0.2420	0.2309	0.2427
Mode	1.0	1.1698	0.9747
Median	2.3660	2.3946	2.3655
20% trimmed mean	2.5049	2.5245	2.5045
Percentiles			
0.01	0.1982	0.1314	0.1984
0.05	0.2892	0.2613	0.2892
0.25	0.4445	0.4397	0.4446
0.75	0.6709	0.6755	0.6709
0.95	0.8071	0.7774	0.8072
0.99	0.8791	0.8371	0.8788

TABLE 3.5

Third- and Fifth-Order Standard Normal Power Method PDF Approximations (Dashed Lines) to χ_6^2 Distribution

$$x = \chi_6^2, \mu = 6.0, \sigma = 3.464102$$

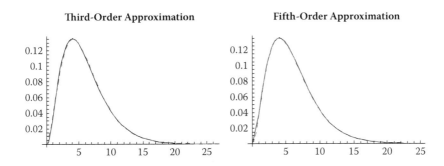

Third-Order Approximation   **Fifth-Order Approximation**

	Standardized Cumulants (γ_i)		Coefficients (c_i)	
x	Third	Fifth	Third	Fifth
$\gamma_1 = 0$	$\gamma_1 = 0$	$\gamma_1 = 0$	$c_1 = -0.183323$	$c_1 = -0.188189$
$\gamma_2 = 1$	$\gamma_2 = 1$	$\gamma_2 = 1$	$c_2 = 0.928875$	$c_2 = 0.934176$
$\gamma_3 = 1.154701$	$\gamma_3 = 1.154701$	$\gamma_3 = 1.154701$	$c_3 = 0.183323$	$c_3 = 0.190397$
$\gamma_4 = 2.0$	$\hat{\gamma}_4 = 2.09$	$\gamma_4 = 2.0$	$c_4 = 0.012158$	$c_4 = 0.009967$
$\gamma_5 = 4.618802$	$\hat{\gamma}_5 = 5.240235$	$\gamma_5 = 4.618802$	$c_5 = 0.0$	$c_5 = -0.000736$
$\gamma_6 = 13.333333$	$\hat{\gamma}_6 = 16.941173$	$\gamma_6 = 13.333333$	$c_6 = 0.0$	$c_6 = 0.000025$

Pdf	x	Third	Fifth
Height	0.1353	0.1350	0.1353
Mode	4.0	4.0925	3.9988
Median	5.3481	5.3469	5.3481
20% trimmed mean	5.4895	5.5012	5.4894
Percentiles			
0.01	0.8721	0.7860	0.8739
0.05	1.6354	1.6030	1.6363
0.25	3.4546	3.4706	3.4543
0.75	7.8408	7.8371	7.8409
0.95	12.5916	12.5632	12.5915
0.99	16.8119	16.8175	16.8119

TABLE 3.6

Third- and Fifth-Order Standard Normal Power Method PDF Approximations
(Dashed Lines) to χ_8^2 Distribution

$$x = \chi_8^2,\ \mu = 8.0,\ \sigma = 4.0$$

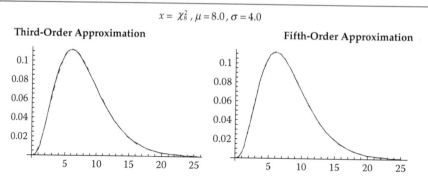

Standardized Cumulants (γ_i)			Coefficients (c_i)	
x	Third	Fifth	Third	Fifth
$\gamma_1 = 0$	$\gamma_1 = 0$	$\gamma_1 = 0$	$c_1 = -0.160481$	$c_1 = -0.163968$
$\gamma_2 = 1$	$\gamma_2 = 1$	$\gamma_2 = 1$	$c_2 = 0.945563$	$c_2 = 0.950794$
$\gamma_3 = 1.0$	$\gamma_3 = 1.0$	$\gamma_3 = 1.0$	$c_3 = 0.160481$	$c_3 = 0.165391$
$\gamma_4 = 1.5$	$\hat{\gamma}_4 = 1.57$	$\gamma_4 = 1.5$	$c_4 = 0.009357$	$c_4 = 0.007345$
$\gamma_5 = 3.0$	$\hat{\gamma}_5 = 3.413916$	$\gamma_5 = 3.0$	$c_5 = 0.0$	$c_5 = -0.000474$
$\gamma_6 = 7.5$	$\hat{\gamma}_6 = 9.566829$	$\gamma_6 = 7.5$	$c_6 = 0.0$	$c_6 = 0.000014$

pdf	x	Third	Fifth
Height	0.1120	0.1121	0.1120
Mode	6.0	6.0752	5.9997
Median	7.3441	7.3581	7.3441
20% trimmed mean	7.4860	7.4958	7.4859
Percentiles			
0.01	1.6465	1.5620	1.6476
0.05	2.7326	2.7070	2.7330
0.25	5.0706	5.0875	5.0705
0.75	10.2189	10.2127	10.2189
0.95	15.5073	15.4827	15.5072
0.99	20.0902	20.1023	20.0902

TABLE 3.7

Third- and Fifth-Order Standard Normal Power Method PDF Approximations (Dashed Lines) to a Weibull(2,5) Distribution

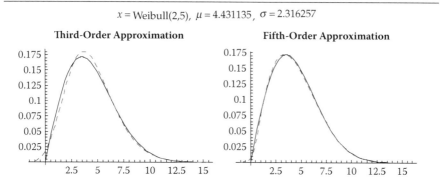

$$x = \text{Weibull(2,5)}, \ \mu = 4.431135, \ \sigma = 2.316257$$

	Standardized Cumulants (γ_i)		Coefficients (c_i)	
x	Third	Fifth	Third	Fifth
$\gamma_1 = 0$	$\gamma_1 = 0$	$\gamma_1 = 0$	$c_1 = -0.103552$	$c_1 = -0.123274$
$\gamma_2 = 1$	$\gamma_2 = 1$	$\gamma_2 = 1$	$c_2 = 0.977678$	$c_2 = 1.044377$
$\gamma_3 = 0.631111$	$\gamma_3 = 0.631111$	$\gamma_3 = 0.631111$	$c_3 = 0.103552$	$c_3 = 0.133015$
$\gamma_4 = 0.245089$	$\hat{\gamma}_4 = 0.625089$	$\gamma_4 = 0.245089$	$c_4 = 0.003832$	$c_4 = -0.028050$
$\gamma_5 = -0.313137$	$\hat{\gamma}_5 = 0.856197$	$\gamma_5 = -0.313137$	$c_5 = 0.0$	$c_5 = -0.003247$
$\gamma_6 = -0.868288$	$\hat{\gamma}_6 = 1.508787$	$\tilde{\gamma}_6 = 0.131712$	$c_6 = 0.0$	$c_6 = 0.001743$

pdf	x	Third	Fifth
Height	0.1716	0.1803	0.1722
Mode	3.5355	3.7120	3.3731
Median	4.1628	4.1913	4.1456
20% trimmed mean	4.2201	4.2428	4.2111
Percentiles			
0.01	0.5013	0.1094	0.5081
0.05	1.1324	1.0759	1.1857
0.25	2.6818	2.7703	2.6720
0.75	5.8870	5.8305	5.8965
0.95	8.6541	8.6046	8.6626
0.99	10.7298	10.8692	10.6773

TABLE 3.8

Third- and Fifth-Order Standard Normal Power Method PDF Approximations (Dashed Lines) to a Weibull(6,5) Distribution

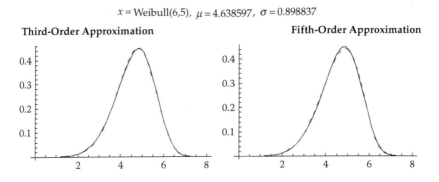

$$x = \text{Weibull}(6,5), \; \mu = 4.638597, \; \sigma = 0.898837$$

Standardized Cumulants (γ_i)			Coefficients (c_i)	
x	Third	Fifth	Third	Fifth
$\gamma_1 = 0$	$\gamma_1 = 0$	$\gamma_1 = 0$	$c_1 = 0.061760$	$c_1 = 0.074862$
$\gamma_2 = 1$	$\gamma_2 = 1$	$\gamma_2 = 1$	$c_2 = 0.991254$	$c_2 = 1.025865$
$\gamma_3 = -0.373262$	$\gamma_3 = -0.373262$	$\gamma_3 = -0.373262$	$c_3 = -0.061760$	$c_3 = -0.085339$
$\gamma_4 = 0.035455$	$\hat{\gamma}_4 = 0.225455$	$\gamma_4 = 0.035455$	$c_4 = 0.001639$	$c_4 = -0.015486$
$\gamma_5 = 0.447065$	$\hat{\gamma}_5 = -0.189751$	$\gamma_5 = 0.447065$	$c_5 = 0.0$	$c_5 = 0.003492$
$\gamma_6 = -1.022066$	$\hat{\gamma}_6 = 0.206084$	$\tilde{\gamma}_6 = -0.022066$	$c_6 = 0.0$	$c_6 = 0.001077$

pdf	x	Third	Fifth
Height	0.4480	0.4513	0.4395
Mode	4.8504	4.8049	4.8764
Median	4.7037	4.6941	4.7059
20% trimmed mean	4.6833	4.6822	4.6897
Percentiles			
0.01	2.3227	2.3024	2.3469
0.05	3.0478	3.0718	3.0549
0.25	4.0625	4.0674	4.0538
0.75	5.2797	5.2703	5.2894
0.95	6.0033	6.0160	5.9877
0.99	6.4493	6.4849	6.4185

TABLE 3.9

Third- and Fifth-Order Standard Normal Power Method PDF Approximations
(Dashed Lines) to a Weibull(10,5) Distribution

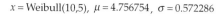

$x = \text{Weibull}(10,5)$, $\mu = 4.756754$, $\sigma = 0.572286$

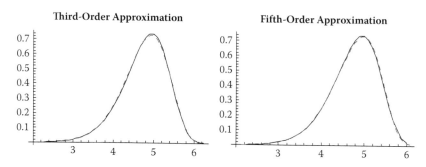

Third-Order Approximation / Fifth-Order Approximation

	Standardized Cumulants (γ_i)		Coefficients (c_i)	
x	Third	Fifth	Third	Fifth
$\gamma_1 = 0$	$\gamma_1 = 0$	$\gamma_1 = 0$	$c_1 = 0.104537$	$c_1 = 0.115914$
$\gamma_2 = 1$	$\gamma_2 = 1$	$\gamma_2 = 1$	$c_2 = 0.976982$	$c_2 = 0.984876$
$\gamma_3 = -0.637637$	$\gamma_3 = -0.637637$	$\gamma_3 = -0.637637$	$c_3 = -0.104537$	$c_3 = -0.126587$
$\gamma_4 = 0.570166$	$\hat{\gamma}_4 = 0.640166$	$\tilde{\gamma}_4 = 0.570166$	$c_4 = 0.003994$	$c_4 = -0.001990$
$\gamma_5 = -0.330588$	$\hat{\gamma}_5 = -0.889751$	$\tilde{\gamma}_5 = -0.330588$	$c_5 = 0.0$	$c_5 = 0.003558$
$\gamma_6 = -0.876003$	$\hat{\gamma}_6 = 1.591787$	$\tilde{\gamma}_6 = 0.123997$	$c_6 = 0.0$	$c_6 = 0.000648$

pdf	x	Third	Fifth
Height	0.7396	0.7304	0.7338
Mode	4.9476	4.9360	4.9734
Median	4.8201	4.8166	4.8231
20% trimmed mean	4.8060	4.8037	4.8077
Percentiles			
0.01	3.1564	3.1633	3.1685
0.05	3.7152	3.7249	3.7155
0.25	4.4143	4.4115	4.4107
0.75	5.1660	5.1672	5.1704
0.95	5.5798	5.5846	5.5685
0.99	5.8250	5.8223	5.8128

TABLE 3.10

Third- and Fifth-Order Standard Normal Power Method PDF Approximations
(Dashed Lines) to a $F_{3,100}$ Distribution

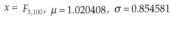

$$x = F_{3,100}, \ \mu = 1.020408, \ \sigma = 0.854581$$

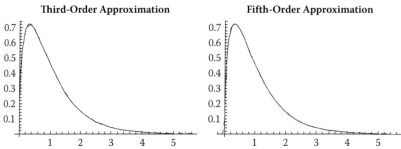

Third-Order Approximation   Fifth-Order Approximation

	Standardized Cumulants (γ_i)		Coefficients (c_i)	
x	Third	Fifth	Third	Fifth
$\gamma_1 = 0$	$\gamma_1 = 0$	$\gamma_1 = 0$	$c_1 = -0.262674$	$c_1 = -0.265108$
$\gamma_2 = 1$	$\gamma_2 = 1$	$\gamma_2 = 1$	$c_2 = 0.842591$	$c_2 = 0.843935$
$\gamma_3 = 1.761428$	$\gamma_3 = 1.761428$	$\gamma_3 = 1.761428$	$c_3 = 0.262674$	$c_3 = 0.267419$
$\gamma_4 = 4.885548$	$\hat{\gamma}_4 = 4.905548$	$\gamma_4 = 4.885548$	$c_4 = 0.027784$	$c_4 = 0.026224$
$\gamma_5 = 18.915442$	$\hat{\gamma}_5 = 7.495308$	$\gamma_5 = 18.915442$	$c_5 = 0.0$	$c_5 = -0.000770$
$\gamma_6 = 95.531520$	$\hat{\gamma}_6 = 24.902004$	$\gamma_6 = 95.531520$	$c_6 = 0.0$	$c_6 = 0.000210$

pdf	*x*	Third	Fifth
Height	0.7224	0.7130	0.7254
Mode	0.3268	0.3369	0.3179
Median	0.7941	0.7959	0.7939
20% trimmed mean	0.8430	0.8441	0.8428
Percentiles			
0.01	0.0381	0.0367	0.0392
0.05	0.1169	0.1132	0.1191
0.25	0.4046	0.4051	0.4043
0.75	1.3909	1.3910	1.3910
0.95	2.6955	2.6933	2.6955
0.99	3.9837	3.9848	3.9835

TABLE 3.11

Third- and Fifth-Order Standard Normal Power Method PDF Approximations
(Dashed Lines) to a $F_{6,100}$ Distribution

$$x = F_{6,100}, \quad \mu = 1.020408, \quad \sigma = 0.613189$$

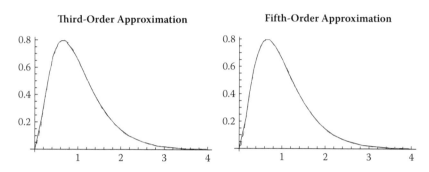

Third-Order Approximation

Fifth-Order Approximation

	Standardized Cumulants (γ_i)		Coefficients (c_i)	
x	Third	Fifth	Third	Fifth
$\gamma_1 = 0$	$\gamma_1 = 0$	$\gamma_1 = 0$	$c_1 = -0.200569$	$c_1 = -0.200690$
$\gamma_2 = 1$	$\gamma_2 = 1$	$\gamma_2 = 1$	$c_2 = 0.907418$	$c_2 = 0.909112$
$\gamma_3 = 1.298235$	$\gamma_3 = 1.298235$	$\gamma_3 = 1.298235$	$c_3 = 0.200569$	$c_3 = 0.199876$
$\gamma_4 = 2.713513$	$\gamma_4 = 2.713513$	$\gamma_4 = 2.713513$	$c_4 = 0.017719$	$c_4 = 0.015651$
$\gamma_5 = 8.051080$	$\hat{\gamma}_5 = 7.928927$	$\gamma_5 = 8.051080$	$c_5 = 0.0$	$c_5 = 0.000272$
$\gamma_6 = 31.528073$	$\hat{\gamma}_6 = 30.028441$	$\gamma_6 = 31.528073$	$c_6 = 0.0$	$c_6 = 0.000106$

pdf	x	Third	Fifth
Height	0.7965	0.7192	0.7965
Mode	0.6536	0.6584	0.6530
Median	0.8974	0.8974	0.8973
20% trimmed mean	0.9237	0.9238	0.9237
Percentiles			
0.01	0.1431	0.1355	0.1430
0.05	0.2694	0.2692	0.2693
0.25	0.5742	0.5758	0.5742
0.75	1.3321	1.3309	1.3321
0.95	2.1906	2.1911	2.1886
0.99	2.9877	2.9905	2.9783

TABLE 3.12

Third- and Fifth-Order Standard Normal Power Method PDF Approximations
(Dashed Lines) to a $F_{9,100}$ Distribution

$$x = F_{9,100}, \quad \mu = 1.020408, \quad \sigma = 0.507837$$

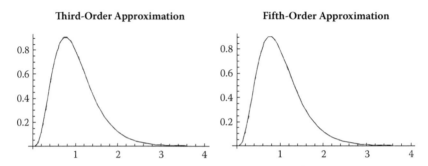

	Standardized Cumulants (γ_i)			Coefficients (c_i)	
x	Third	Fifth		Third	Fifth
$\gamma_1 = 0$	$\gamma_1 = 0$	$\gamma_1 = 0$	$c_1 = -0.172243$	$c_1 = -0.171699$
$\gamma_2 = 1$	$\gamma_2 = 1$	$\gamma_2 = 1$	$c_2 = 0.926083$	$c_2 = 0.930540$
$\gamma_3 = 1.102041$	$\gamma_3 = 1.102041$	$\gamma_3 = 1.102041$	$c_3 = 0.172243$	$c_3 = 0.170286$
$\gamma_4 = 1.991778$	$\gamma_4 = 1.991778$	$\gamma_4 = 1.991778$	$c_4 = 0.014385$	$c_4 = 0.012439$
$\gamma_5 = 5.172909$	$\hat{\gamma}_5 = 5.048491$	$\gamma_5 = 5.172909$	$c_5 = 0.0$	$c_5 = 0.000471$
$\gamma_6 = 17.862380$	$\hat{\gamma}_6 = 16.603145$	$\gamma_6 = 17.862380$	$c_6 = 0.0$	$c_6 = 0.000079$

pdf	x	Third	Fifth
Height	0.9064	0.9120	0.9063
Mode	0.7625	0.7627	0.7623
Median	0.9332	0.9329	0.9332
20% trimmed mean	0.9518	0.9517	0.9518
Percentiles			
0.01	0.2265	0.2203	0.2267
0.05	0.3629	0.3635	0.3630
0.25	0.6520	0.6533	0.6519
0.75	1.2933	1.2922	1.2933
0.95	1.9748	1.9757	1.9748
0.99	2.5898	2.5924	2.5898

TABLE 3.13

Third- and Fifth-Order Standard Normal Power Method PDF Approximations
(Dashed Lines) to a t_7 Distribution

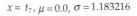

$$x = t_7, \mu = 0.0, \sigma = 1.183216$$

Third-Order Approximation   **Fifth-Order Approximation**

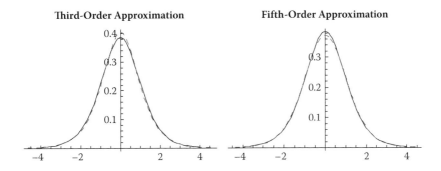

	Standardized Cumulants (γ_i)			Coefficients (c_i)	
x	Third	Fifth		Third	Fifth
$\gamma_1 = 0$	$\gamma_1 = 0$	$\gamma_1 = 0$		$c_1 = 0.0$	$c_1 = 0.0$
$\gamma_2 = 1$	$\gamma_2 = 1$	$\gamma_2 = 1$		$c_2 = 0.835665$	$c_2 = 0.907394$
$\gamma_3 = 0.0$	$\gamma_3 = 0.0$	$\gamma_3 = 0.0$		$c_3 = 0.0$	$c_3 = 0.0$
$\gamma_4 = 2.0$	$\gamma_4 = 2.0$	$\gamma_4 = 2.0$		$c_4 = 0.052057$	$c_4 = 0.014980$
$\gamma_5 = 0.0$	$\hat{\gamma}_5 = 0.0$	$\gamma_5 = 0.0$		$c_5 = 0.0$	$c_5 = 0.0$
$\gamma_6 = 80.0$	$\hat{\gamma}_6 = 24.289447$	$\gamma_6 = 80.0$		$c_6 = 0.0$	$c_6 = 0.002780$

pdf	x	Third	Fifth
Height	0.3850	0.4035	0.3716
Mode	0.0	0.0	0.0
Median	0.0	0.0	0.0
20% trimmed mean	0.0	0.0	0.0
Percentiles			
0.01	-2.9979	-3.0757	-2.9449
0.05	-1.8946	-1.9005	-1.8845
0.25	-0.7111	-0.6858	-0.7301
0.75	0.7111	0.6858	0.7301
0.95	1.8946	1.9005	1.8845
0.99	2.9979	3.0757	2.9449

TABLE 3.14

Third- and Fifth-Order Standard Normal Power Method PDF Approximations
(Dashed Lines) to a t_{10} Distribution

$$x = t_{10}, \ \mu = 0.0, \ \sigma = 1.118034$$

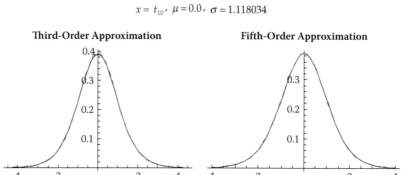

| Third-Order Approximation | Fifth-Order Approximation |

	Standardized Cumulants (γ_i)		Coefficients (c_i)	
x	Third	Fifth	Third	Fifth
$\gamma_1 = 0$	$\gamma_1 = 0$	$\gamma_1 = 0$	$c_1 = 0.0$	$c_1 = 0.0$
$\gamma_2 = 1$	$\gamma_2 = 1$	$\gamma_2 = 1$	$c_2 = 0.902977$	$c_2 = 0.920482$
$\gamma_3 = 0.0$	$\gamma_3 = 0.0$	$\gamma_3 = 0.0$	$c_3 = 0.0$	$c_3 = 0.0$
$\gamma_4 = 1.0$	$\gamma_4 = 1.0$	$\gamma_4 = 1.0$	$c_4 = 0.031356$	$c_4 = 0.021453$
$\gamma_5 = 0.0$	$\hat{\gamma}_5 = 0.0$	$\gamma_5 = 0.0$	$c_5 = 0.0$	$c_5 = 0.0$
$\gamma_6 = 10.0$	$\hat{\gamma}_6 = 5.938334$	$\gamma_6 = 10.0$	$c_6 = 0.0$	$c_6 = 0.000830$

pdf	x	Third	Fifth
Height	0.3891	0.3952	0.3876
Mode	0.0	0.0	0.0
Median	0.0	0.0	0.0
20% trimmed mean	0.0	0.0	0.0
Percentiles			
0.01	−2.7638	−2.7900	−2.6961
0.05	−1.8125	−1.8166	−1.7995
0.25	−0.6998	−0.6917	−0.7015
0.75	0.6998	0.6917	0.7015
0.95	1.8125	1.8166	1.7995
0.99	2.7638	2.7900	2.6961

TABLE 3.15

Third- and Fifth-Order Standard Normal Power Method PDF Approximations (Dashed Lines) to a t_{20} Distribution

$x = t_{20},\ \mu = 0.0,\ \sigma = 1.054093$

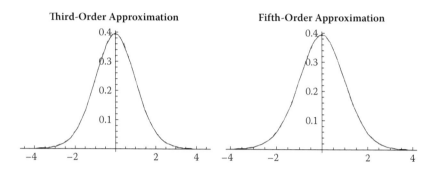

Standardized Cumulants (γ_i)			Coefficients (c_i)	
x	Third	Fifth	Third	Fifth
$\gamma_1 = 0$	$\gamma_1 = 0$	$\gamma_1 = 0$	$c_1 = 0.0$	$c_1 = 0.0$
$\gamma_2 = 1$	$\gamma_2 = 1$	$\gamma_2 = 1$	$c_2 = 0.958053$	$c_2 = 0.960799$
$\gamma_3 = 0.0$	$\gamma_3 = 0.0$	$\gamma_3 = 0.0$	$c_3 = 0.0$	$c_3 = 0.0$
$\gamma_4 = 0.375$	$\gamma_4 = 0.375$	$\gamma_4 = 0.375$	$c_4 = 0.013792$	$c_4 = 0.012099$
$\gamma_5 = 0.0$	$\hat{\gamma}_5 = 0.0$	$\gamma_5 = 0.0$	$c_5 = 0.0$	$c_5 = 0.0$
$\gamma_6 = 1.071429$	$\hat{\gamma}_6 = 0.812082$	$\gamma_6 = 1.071429$	$c_6 = 0.0$	$c_6 = 0.000156$

pdf	x	Third	Fifth
Height	0.3940	0.3950	0.3939
Mode	0.0	0.0	0.0
Median	0.0	0.0	0.0
20% trimmed mean	0.0	0.0	0.0
Percentiles			
0.01	−2.5280	−2.5232	−2.5166
0.05	−1.7247	−1.7258	−1.7226
0.25	−0.6870	−0.6856	−0.6870
0.75	0.6870	0.6856	0.6870
0.95	1.7247	1.7258	1.7226
0.99	2.5280	2.5232	2.5166

fifth-order system and subsequently evaluating. For example, the values of γ_5 and γ_6 for the third-order approximations to the beta distributions are quite different from the exact values of γ_5 and γ_6 for the fifth-order approximations. In contrast, in the context of the chi-square distributions, the values of γ_5 and γ_6 associated with the third-order polynomials are much closer to the exact (or fifth-order polynomial) cumulants and thus provide more accurate approximations to these theoretical pdfs.

In terms of empirical pdfs, presented in Tables 3.16–3.21 are fifth-order power method pdfs superimposed on measures taken of body density, fat, forearm, thigh, abdomen, and height from $N = 252$ adult males (http://lib. stat.cmu.edu/datasets/bodyfat).

Similar to the theoretical pdfs previously discussed, these examples demonstrate the usefulness of the software for making comparisons in terms of computing cumulants and other various measures of central tendency. One way of determining how well the power method pdfs model these data is to compute chi-square goodness of fit statistics. For example, listed in Tables 3.16–3.21 are the cumulative percentages and class intervals based on the power method pdfs. The asymptotic p-values indicate the power method pdfs provide fairly good fits to the data. It is noted that the degrees of freedom for these tests were computed as $df = 3 = 10$ (class intervals) – 6 (parameter estimates) – 1 (sample size).

Presented in Tables 3.22–3.38 are examples of symmetric and asymmetric logistic and uniform power method pdfs. The calculations for the coefficients, measures of central tendency, percentiles, graphing, and so forth, were accomplished by adapting the source code provided by Headrick et al. (2007) using the methodology developed for the logistic and uniform distributions in the previous chapter. The source code for both transformations has the *Mathematica* functions listed in Table 2.4 and is available from the author.

One of the purposes of presenting Tables 3.22–3.38 is to provide a guide for the user in terms of the cumulants. Specifically, these cumulants can act as initial starting points for a user who may want a distribution with similar cumulants. And, another point of interest associated with these tables is that they illustrate how the shapes of some distributions can be markedly different even though some distributions have the same values of skew and kurtosis. For example, the uniform-based third- and fifth-order pdfs in Table 3.32 are unimodal and bimodal distributions, respectively, while the only difference in terms of the cumulants is the value of γ_6 is slightly larger for the fifth-order polynomial. Further, the third-order pdf in Table 3.33 is much more peaked than the fifth-order pdf, where the only difference is again that the sixth cumulant is slightly larger. Similarly, in the context of the logistic polynomials, depicted in Table 3.23 are symmetric third- and fifth-order pdfs where the values of the first five cumulants are the same. However, the third-order pdf is much more peaked than the fifth-order pdf, where the effect of increasing the six cumulants has the effect of reducing the height of the pdf.

TABLE 3.16

Fifth-Order Normal-Based Power Method (PM) Approximation and Goodness of Fit Test for the Body Density Data Taken from $N = 252$ Men

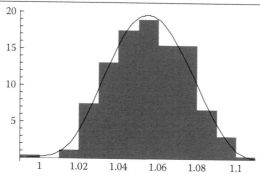

	Data	PM Distribution	PM Coefficients
Mean	1.05557	1.05557	$c_1 = -0.006154$
Standard deviation	0.01903	0.01903	$c_2 = 1.071691$
Skew	−0.02018	−0.02018	$c_3 = 0.016487$
Kurtosis	−0.30962	−0.30962	$c_4 = -0.033401$
Fifth cumulant	−0.40038	−0.40038	$c_5 = -0.003444$
Sixth cumulant	2.36824	2.36824	$c_6 = 0.001831$
Mode(s)	1.061	1.055	
First quartile	1.041	1.042	
Median	1.055	1.055	
Third quartile	1.070	1.069	
20% TM	1.055	1.055	

Cumulative %	PM Class Intervals	Observed Data Freq	Expected Freq
10	< 1.03087	25	25.2
20	1.03087–1.03885	27	25.2
30	1.03885–1.04495	24	25.2
40	1.04495–1.05034	28	25.2
50	1.05034–1.05548	25	25.2
60	1.05548–1.06065	22	25.2
70	1.06065–1.06615	21	25.2
80	1.06615–1.07243	27	25.2
90	1.07243–1.08065	28	25.2
100	> 1.08065	25	25.2

$\chi^2 = 2.048$ $\Pr\{\chi_9^2 \leq 2.048\} = 0.438$

TABLE 3.17

Fifth-Order Normal-Based Power Method (PM) Approximation and Goodness of Fit Test for the Body Fat Data Taken from $N = 252$ Men

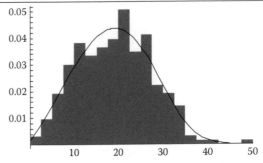

	Data	PM Distribution	PM Coefficients
Mean	19.1508	19.1508	$c_1 = -0.012608$
Standard deviation	8.36874	8.36874	$c_2 = 1.093208$
Skew	0.14635	0.14635	$c_3 = -0.003880$
Kurtosis	-0.33381	-0.33381	$c_4 = -0.042950$
Fifth cumulant	0.53391	0.53391	$c_5 = 0.005496$
Sixth cumulant	3.01832	5.00000	$c_6 = 0.002187$
Mode(s)	20.40	19.13	
First quartile	12.42	12.98	
Median	19.20	19.04	
Third quartile	25.30	25.10	
20% TM	19.07	19.04	

Cumulative %	PM Class Intervals	Observed Data Freq	Expected Freq
10	< 8.08465	25	25.2
20	8.08465–11.5521	28	25.2
30	11.5521–14.2933	25	25.2
40	14.2933–16.7314	21	25.2
50	16.7314–19.0453	24	25.2
60	19.0453–21.3554	28	25.2
70	21.3554–23.7864	25	25.2
80	23.7864–26.5386	24	25.2
90	26.5386–30.1474	27	25.2
100	> 30.1474	25	25.2

$\chi^2 = 1.571$ $\Pr\{\chi_3^2 \leq 1.571\} = 0.334$

TABLE 3.18

Fifth-Order Normal-Based Power Method (PM) Approximation and Goodness of Fit Test for the Forearm Data Taken from $N = 252$ Men

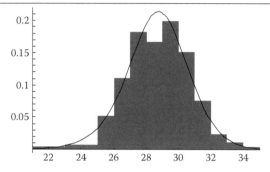

	Data	PM Distribution	PM Coefficients
Mean	28.6639	28.6639	$c_1 = 0.023244$
Standard deviation	2.02069	2.02069	$c_2 = 0.924804$
Skew	−0.21933	−0.21933	$c_3 = -0.016335$
Kurtosis	0.86631	0.86631	$c_4 = 0.023136$
Fifth cumulant	−1.61834	−1.61834	$c_5 = -0.002303$
Sixth cumulant	3.16298	7.00000	$c_6 = 0.000195$
Mode(s)	27.30, 29.60, 29.80	28.77	
First quartile	27.30	27.42	
Median	28.70	28.71	
Third quartile	30.00	29.97	
20% TM	28.70	28.70	

Cumulative %	PM Class Intervals	Observed Data Freq	Expected Freq
10	< 26.1494	23	25.2
20	26.1494–27.0843	28	25.2
30	27.0843–27.7147	28	25.2
40	27.7147–28.2345	25	25.2
50	28.2345–28.7109	23	25.2
60	28.7109–29.1829	22	25.2
70	29.1829–29.6882	24	25.2
80	29.6882–30.2859	27	25.2
90	30.2859–31.1388	30	25.2
100	> 31.1388	22	25.2

$\chi^2 = 2.921$ $\Pr\{\chi_3^2 \leq 2.921\} = 0.596$

TABLE 3.19

Fifth-Order Normal-Based Power Method (PM) Approximation and Goodness of
Fit Test for the Thigh Data Taken from $N = 252$ Men

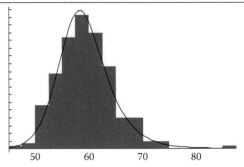

	Data	PM Distribution	PM Coefficients
Mean	59.4059	59.4059	$c_1 = -0.086727$
Standard deviation	5.24995	5.24995	$c_2 = 0.862343$
Skew	0.82121	0.82121	$c_3 = 0.070869$
Kurtosis	2.66571	2.66571	$c_4 = 0.039130$
Fifth cumulant	11.3881	11.3881	$c_5 = 0.005286$
Sixth cumulant	47.4639	72.0000	$c_6 = 0.000268$
Mode(s)	58.90	58.36	
First quartile	56.00	56.01	
Median	59.00	58.95	
Third quartile	62.45	62.25	
20% TM	59.13	59.03	

Cumulative %	PM Class Intervals	Observed Data Freq	Expected Freq
10	< 53.3974	28	25.2
20	53.3974–55.2948	24	25.2
30	55.2948–56.6513	21	25.2
40	56.6513–57.8243	28	25.2
50	57.8243–58.9506	24	25.2
60	58.9506–60.1250	25	25.2
70	60.1250–61.4588	26	25.2
80	61.4588–63.1614	20	25.2
90	63.1614–65.8757	30	25.2
100	> 65.8757	26	25.2

$\chi^2 = 3.476$ $\qquad$ $\Pr\{\chi^2_3 \le 3.476\} = 0.676$

TABLE 3.20

Fifth-Order Normal-Based Power Method (PM) Approximation and Goodness of Fit Test for the Abdomen Data Taken from $N = 252$ Men

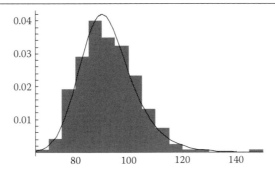

	Data	PM Distribution	PM Coefficients
Mean	92.5560	92.5560	$c_1 = -0.095884$
Standard deviation	10.7831	10.7831	$c_2 = 0.896280$
Skew	0.83842	0.83842	$c_3 = 0.079266$
Kurtosis	2.24882	2.24882	$c_4 = 0.027407$
Fifth cumulant	9.30623	9.30623	$c_5 = 0.005539$
Sixth cumulant	37.4539	55.0000	$c_6 = 0.000363$
Mode(s)	88.70, 89.70, 100.5	90.05	
First quartile	84.53	82.81	
Median	90.95	89.02	
Third quartile	99.58	96.03	
20% TM	91.81	89.21	

Cumulative %	PM Class Intervals	Observed Data Freq	Expected Freq
10	< 80.0656	30	25.2
20	80.0656–83.8456	27	25.2
30	83.8456–86.6507	20	25.2
40	86.6507–89.1238	24	25.2
50	89.1238–91.5220	28	25.2
60	91.5220–94.0304	20	25.2
70	94.0304–96.8725	23	25.2
80	96.8725–100.469	27	25.2
90	100.469–106.108	30	25.2
100	> 106.108	23	25.2

$\chi^2 = 4.984$ $\qquad$ $\Pr\{\chi_3^2 \leq 4.984\} = 0.827$

TABLE 3.21

Fifth-Order Normal-Based Power Method (PM) Approximation and Goodness of Fit for the Height Data Taken from $N = 252$ Men

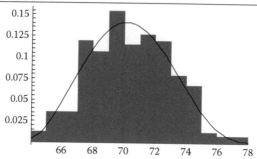

	Data	PM Distribution	PM Coefficients
Mean	70.3075	70.3075	$c_1 = -0.017943$
Standard deviation	2.60958	2.60958	$c_2 = 1.086544$
Skew	0.10262	0.10262	$c_3 = 0.016367$
Kurtosis	-0.40392	-0.40392	$c_4 = -0.038981$
Fifth cumulant	-0.09502	-0.09502	$c_5 = 0.000525$
Sixth cumulant	1.82126	1.82126	$c_6 = 0.001907$
Mode(s)	71.50	70.15	
First quartile	68.25	68.39	
Median	70.00	70.26	
Third quartile	72.25	72.16	
20% TM	70.26	70.27	

Cumulative %	PM Class Intervals	Observed Data Freq	Expected Freq
10	< 66.8977	21	25.2
20	66.8977–67.9638	30	25.2
30	67.9638–68.8001	27	25.2
40	68.8001–69.5468	30	25.2
50	69.5468–70.2607	25	25.2
60	70.2607–70.9802	18	25.2
70	70.9802–71.7450	27	25.2
80	71.7450–72.6195	24	25.2
90	72.6195–73.7714	26	25.2
100	> 73.7714	24	25.2

$\chi^2 = 4.984$ $\Pr\{\chi_3^2 \le 4.984\} = 0.827$

TABLE 3.22

Third- and Fifth-Order Symmetric Logistic-Based Power Method PDFs and Their Associated Cumulants, Coefficients, and Other Indices

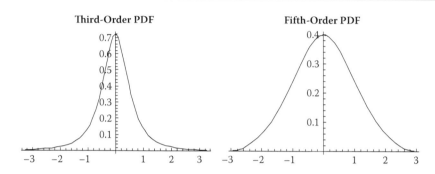

Cumulants	Coefficients	Cumulants	Coefficients
$\gamma_1 = 0$	$c_1 = 0.0$	$\gamma_1 = 0$	$c_1 = 0.0$
$\gamma_2 = 1$	$c_2 = 0.630162$	$\gamma_2 = 1$	$c_2 = 1.136629$
$\gamma_3 = 0$	$c_3 = 0.0$	$\gamma_3 = 0$	$c_3 = 0.0$
$\gamma_4 = 50$	$c_4 = 0.073367$	$\gamma_4 = 50$	$c_4 = -0.047708$
$\gamma_5 = 0$	$c_5 = 0.0$	$\gamma_5 = 0$	$c_5 = 0.0$
$\gamma_6 = 76956.946$	$c_6 = 0.0$	$\gamma_6 = 35927500$	$c_6 = 0.001410$

Height	0.719577	Height	0.398943
Mode	0.0	Mode	0.0
Median	0.0	Median	0.0
20% trimmed mean	0.0	20% trimmed mean	0.0

Percentiles		*Percentiles*	
0.01	−2.7894	0.01	−2.2509
0.05	−1.8774	0.05	−1.6569
0.25	−0.3880	0.25	−0.6779
0.75	0.3880	0.75	0.6779
0.95	1.8774	0.95	1.6569
0.99	2.7894	0.99	2.2509

TABLE 3.23

Third- and Fifth-Order Symmetric Logistic-Based Power Method PDFs and Their
Associated Cumulants, Coefficients, and Other Indices

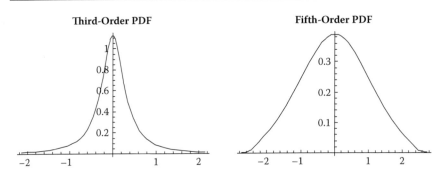

Cumulants	Coefficients	Cumulants	Coefficients
$\gamma_1 = 0$	$c_1 = 0.0$	$\gamma_1 = 0$	$c_1 = 0.0$
$\gamma_2 = 1$	$c_2 = 0.402998$	$\gamma_2 = 1$	$c_2 = 1.164415$
$\gamma_3 = 0$	$c_3 = 0.0$	$\gamma_3 = 0$	$c_3 = 0.0$
$\gamma_4 = 150$	$c_4 = 0.108591$	$\gamma_4 = 150$	$c_4 = -0.0595505$
$\gamma_5 = 0$	$c_5 = 0.0$	$\gamma_5 = 0$	$c_5 = 0.0$
$\gamma_6 = 589653.8566$	$c_6 = 0.0$	$\gamma_6 = 190000000$	$c_6 = 0.0018532$

Height	1.125190	Height	0.389423
Mode	0.0	Mode	0.0
Median	0.0	Median	0.0
20% trimmed mean	0.0	20% trimmed mean	0.0

Percentiles		*Percentiles*	
0.01	−2.7866	0.01	−2.1750
0.05	−1.1188	0.05	−1.6564
0.25	−0.2682	0.25	−0.6922
0.75	0.2682	0.75	0.6922
0.95	1.1188	0.95	1.6564
0.99	2.7866	0.99	2.1750

TABLE 3.24

Third- and Fifth-Order Asymmetric Logistic-Based Power Method PDFs and Their Associated Cumulants, Coefficients, and Other Indices

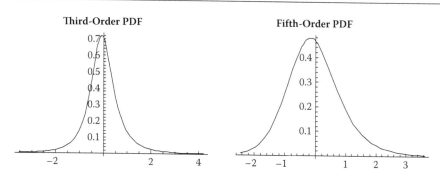

Cumulants	Coefficients	Cumulants	Coefficients
$\gamma_1 = 0$	$c_1 = -0.041855$	$\gamma_1 = 0$	$c_1 = -0.040785$
$\gamma_2 = 1$	$c_2 = 0.632103$	$\gamma_2 = 1$	$c_2 = 0.772780$
$\gamma_3 = 1$	$c_3 = 0.041855$	$\gamma_3 = 1$	$c_3 = 0.038698$
$\gamma_4 = 50$	$c_4 = 0.072535$	$\gamma_4 = 50$	$c_4 = 0.041112$
$\gamma_5 = 439.4782$	$c_5 = 0.0$	$\gamma_5 = 1500$	$c_5 = 0.000497$
$\gamma_6 = 75884.3015$	$c_6 = 0.0$	$\gamma_6 = 10000000$	$c_6 = 0.0004885$
Height	0.720084	Height	0.588284
Mode	−0.077793	Mode	−0.080279
Median	−0.041855	Median	−0.040785
20% trimmed mean	−0.034642	20% trimmed mean	−0.034088
Percentiles		*Percentiles*	
0.01	−2.5540	0.01	−2.4493
0.05	−1.2680	0.05	−1.3713
0.25	−0.4255	0.25	−0.5037
0.75	0.3725	0.75	0.3085
0.95	1.4049	0.95	1.1504
0.99	3.0076	0.99	2.6846

TABLE 3.25

Third- and Fifth-Order Asymmetric Logistic-Based Power Method PDFs and Their Associated Cumulants, Coefficients, and Other Indices

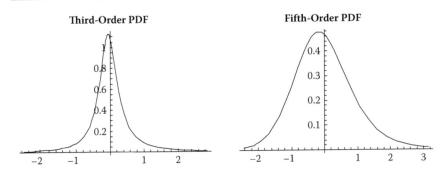

Cumulants	Coefficients	Cumulants	Coefficients
$\gamma_1 = 0$	$c_1 = -0.029705$	$\gamma_1 = 0$	$c_1 = -0.072807$
$\gamma_2 = 1$	$c_2 = 0.403351$	$\gamma_2 = 1$	$c_2 = 0.959758$
$\gamma_3 = 1$	$c_3 = 0.029705$	$\gamma_3 = 1$	$c_3 = 0.074581$
$\gamma_4 = 150$	$c_4 = 0.108306$	$\gamma_4 = 150$	$c_4 = -0.088068$
$\gamma_5 = 1089.27783$	$c_5 = 0.0$	$\gamma_5 = 500$	$c_5 = -0.000422$
$\gamma_6 = 587501.083$	$c_6 = 0.0$	$\gamma_6 = 75000000$	$c_6 = 0.001434$
Height	1.127973	Height	0.476128
Mode	−0.047951	Mode	−0.167651
Median	−0.029705	Median	−0.072807
20% trimmed mean	−0.024586	20% trimmed mean	−0.059978
Percentiles		*Percentiles*	
0.01	−2.6225	0.01	−2.0494
0.05	−1.0695	0.05	−1.4158
0.25	−0.2871	0.25	−0.6249
0.75	0.2496	0.75	0.5340
0.95	1.1667	0.95	1.6574
0.99	2.9439	0.99	2.8265

TABLE 3.26

Third- and Fifth-Order Asymmetric Logistic-Based Power Method PDFs and Their Associated Cumulants, Coefficients, and Other Indices

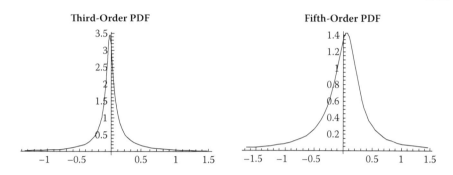

Cumulants	Coefficients	Cumulants	Coefficients
$\gamma_1 = 0$	$c_1 = -0.022634$	$\gamma_1 = 0$	$c_1 = 0.034014$
$\gamma_2 = 1$	$c_2 = 0.132247$	$\gamma_2 = 1$	$c_2 = 0.324360$
$\gamma_3 = 1$	$c_3 = 0.022634$	$\gamma_3 = 1$	$c_3 = -0.044329$
$\gamma_4 = 350$	$c_4 = 0.143557$	$\gamma_4 = 350$	$c_4 = 0.112698$
$\gamma_5 = 2099.5227$	$c_5 = 0.0$	$\gamma_5 = 20000$	$c_5 = 0.002456$
$\gamma_6 = 2639787.896$	$c_6 = 0.0$	$\gamma_6 = 10000000$	$c_6 = 0.000419$
Height	3.453660	Height	1.412194
Mode	−0.028164	Mode	0.057714
Median	−0.022634	Median	0.034014
20% trimmed mean	−0.018734	20% trimmed mean	0.026515
Percentiles		*Percentiles*	
0.01	−2.5465	0.01	−2.8473
0.05	−0.7918	0.05	−1.0791
0.25	−0.1263	0.25	−0.2034
0.75	0.0977	0.75	0.2396
0.95	0.8659	0.95	0.9477
0.99	2.7920	0.99	2.5487

TABLE 3.27

Third- and Fifth-Order Asymmetric Logistic-Based Power Method PDFs and Their Associated Cumulants, Coefficients, and Other Indices

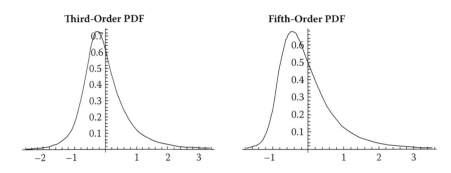

Third-Order PDF		Fifth-Order PDF	
Cumulants	Coefficients	Cumulants	Coefficients
$\gamma_1 = 0$	$c_1 = -0.138784$	$\gamma_1 = 0$	$c_1 = -0.207153$
$\gamma_2 = 1$	$c_2 = 0.648952$	$\gamma_2 = 1$	$c_2 = 0.743320$
$\gamma_3 = 3$	$c_3 = 0.138784$	$\gamma_3 = 3$	$c_3 = 0.216940$
$\gamma_4 = 50$	$c_4 = 0.064482$	$\gamma_4 = 50$	$c_4 = 0.035374$
$\gamma_5 = 1109.249$	$c_5 = 0.0$	$\gamma_5 = -500$	$c_5 = -0.002330$
$\gamma_6 = 65435.531$	$c_6 = 0.0$	$\gamma_6 = 1000000$	$c_6 = 0.000500$
Height	0.730001	Height	0.679113
Mode	−0.265679	Mode	−0.454470
Median	−0.138784	Median	−0.207153
20% trimmed mean	−0.114867	20% trimmed mean	−0.169900
Percentiles		*Percentiles*	
0.01	−1.9405	0.01	−1.4212
0.05	−1.1023	0.05	−1.0153
0.25	−0.4952	0.25	−0.5860
0.75	0.3196	0.75	0.3303
0.95	1.5563	0.95	1.7120
0.99	3.4445	0.99	3.5997

TABLE 3.28

Third- and Fifth-Order Asymmetric Logistic-Based Power Method PDFs and Their Associated Cumulants, Coefficients, and Other Indices

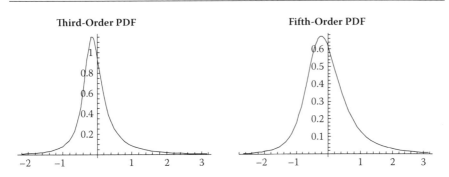

Cumulants	Coefficients	Cumulants	Coefficients
$\gamma_1 = 0$	$c_1 = -0.091565$	$\gamma_1 = 0$	$c_1 = -0.110345$
$\gamma_2 = 1$	$c_2 = 0.406064$	$\gamma_2 = 1$	$c_2 = 0.688447$
$\gamma_3 = 3$	$c_3 = 0.091565$	$\gamma_3 = 3$	$c_3 = 0.107905$
$\gamma_4 = 150$	$c_4 = 0.105906$	$\gamma_4 = 150$	$c_4 = 0.048818$
$\gamma_5 = 3157.8032$	$c_5 = 0.0$	$\gamma_5 = 10000$	$c_5 = 0.000581$
$\gamma_6 = 569447.856$	$c_6 = 0.0$	$\gamma_6 = 10000000$	$c_6 = 0.000782$
Height	1.154089	Height	0.675177
Mode	-0.148657	Mode	-0.216513
Median	-0.091565	Median	-0.110345
20% trimmed mean	-0.075785	20% trimmed mean	-0.0917165
Percentiles		*Percentiles*	
0.01	-2.2546	0.01	-2.0134
0.05	-0.9625	0.05	-1.1572
0.25	-0.3274	0.25	-0.4986
0.75	0.2116	0.75	0.3572
0.95	1.2620	0.95	1.5133
0.99	3.2469	0.99	3.2257

TABLE 3.29

Third- and Fifth-Order Asymmetric Logistic-Based Power Method PDFs and Their Associated Cumulants, Coefficients, and Other Indices

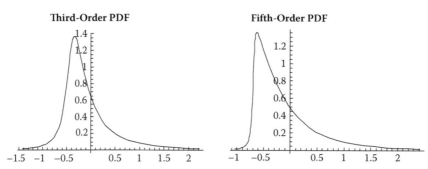

Third-Order PDF			Fifth-Order PDF		
Cumulants	**Coefficients**		**Cumulants**	**Coefficients**	
$\gamma_1 = 0$	$c_1 = -0.207332$		$\gamma_1 = 0$	$c_1 = -0.280605$	
$\gamma_2 = 1$	$c_2 = 0.411615$		$\gamma_2 = 1$	$c_2 = 0.565434$	
$\gamma_3 = 6$	$c_3 = 0.207332$		$\gamma_3 = 6$	$c_3 = 0.283102$	
$\gamma_4 = 150$	$c_4 = 0.095438$		$\gamma_4 = 150$	$c_4 = 0.052509$	
$\gamma_5 = 5457.6591$	$c_5 = 0.0$		$\gamma_5 = 8000$	$c_5 = -0.000594$	
$\gamma_6 = 493071.9367$	$c_6 = 0.0$		$\gamma_6 = 3500000$	$c_6 = 0.000595$	
Height	1.367306		Height	1.358166	
Mode	−0.345501		Mode	−0.632463	
Median	−0.207332		Median	−0.280605	
20% trimmed mean	−0.171602		20% trimmed mean	−0.231852	
Percentiles			*Percentiles*		
0.01	−1.4712		0.01	−0.8364	
0.05	−0.7375		0.05	−0.6879	
0.25	−0.4018		0.25	−0.5311	
0.75	0.1393		0.75	0.1774	
0.95	1.4155		0.95	1.6106	
0.99	3.7180		0.99	3.8603	

TABLE 3.30

Fifth-Order Logistic-Based Power Method PDFs Approximations of Two $t_{df=3}$ Distributions

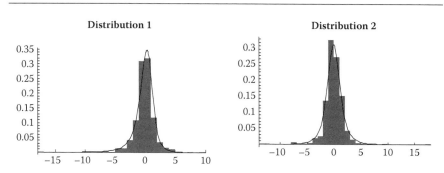

Cumulants	Coefficients	Cumulants	Coefficients
$\mu = -0.04400$	$c_1 = 0.114676$	$\mu = -0.008050$	$c_1 = -0.006886$
$\sigma^2 = 3.19010$	$c_2 = 0.754070$	$\sigma^2 = 3.93140$	$c_2 = 0.753854$
$\gamma_3 = -1.7339$	$c_3 = -0.119800$	$\gamma_3 = 0.168484$	$c_3 = 0.006320$
$\gamma_4 = 21.9084$	$c_4 = 0.047802$	$\gamma_4 = 21.1908$	$c_4 = 0.051467$
$\gamma_5 = -166.3340$	$c_5 = 0.001220$	$\gamma_5 = 56.2067$	$c_5 = 0.000135$
$\hat{\gamma}_6 = 12087.30^1$	$c_6 = 0.000008$	$\hat{\gamma}_6 = 13841.80^2$	$c_6 = 1.208\text{E}-07$

Note: Samples of size $N = 500$ were drawn from each of the two populations.

[1,2] The actual values of the sixth cumulant computed on the data sets were $\gamma_6 = 1677.30$ and $\gamma_6 = 1096.80$ for Distributions 1 and 2, respectively.

The logistic-based polynomials are perhaps most useful when values of kurtosis associated with data sets are large. Often times, if normal-based polynomials are used to approximate data with large values of kurtosis, then excessively peaked power method pdfs will result. For example, consider the two histograms of data drawn from a $t_{df=3}$ distribution in Table 3.30. For these data, logistic-based polynomials will provide much more accurate approximations than normal-based polynomials.

The uniform-based polynomials can often be used to model multinomial data such as those distributions in Table 3.37. Oftentimes continuous variables are treated as ordered categorical data. For example, to increase the probability of response on surveys, variables such as annual income, age, years experience, respondents are asked to provide an integer value that represents their income in some range, such as score $1 < 25K$, $25K \le 2 < 50K$, and so forth, rather than reporting their actual annual income. The uniform-based polynomials have an advantage in this context because the user can specify a power method's pdf to correspond with the real lower (*low*) and upper (*upp*) limits of the multinomial distribution. This can be accomplished by

TABLE 3.31

Third- and Fifth-Order Symmetric Uniform-Based Power Method PDFs and Their
Associated Cumulants, Coefficients, and Other Indices

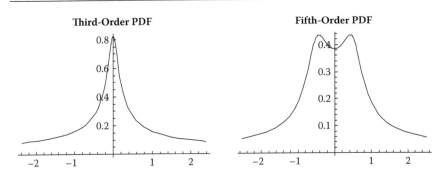

Cumulants	Coefficients	Cumulants	Coefficients
$\gamma_1 = 0$	$c_1 = 0.0$	$\gamma_1 = 0$	$c_1 = 0.0$
$\gamma_2 = 1$	$c_2 = 0.350940$	$\gamma_2 = 1$	$c_2 = 0.750739$
$\gamma_3 = 0$	$c_3 = 0.0$	$\gamma_3 = 0$	$c_3 = 0.0$
$\gamma_4 = 0$	$c_4 = 0.340361$	$\gamma_4 = 0$	$c_4 = -0.161706$
$\gamma_5 = 0$	$c_5 = 0.0$	$\gamma_5 = 0$	$c_5 = 0.0$
$\gamma_6 = -3.464097$	$c_6 = 0.0$	$\gamma_6 = -2.50$	$c_6 = 0.132848$

Height	0.822577	Height	0.435957
Mode	0.0	Mode	±0.428687
Median	0.0	Median	0.0
20% trimmed mean	0.0	20% trimmed mean	0.0

Percentiles		*Percentiles*	
0.01	−2.2601	0.01	−2.3554
0.05	−1.8363	0.05	−1.7805
0.25	−0.5250	0.25	−0.6098
0.75	0.5250	0.75	0.6098
0.95	1.8363	0.95	1.7805
0.99	2.2601	0.99	2.3554

TABLE 3.32

Third- and Fifth-Order Symmetric Uniform-Based Power Method PDFs and Their Associated Cumulants, Coefficients, and Other Indices

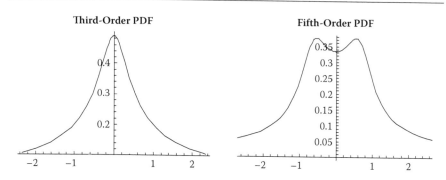

Cumulants	Coefficients	Cumulants	Coefficients
$\gamma_1 = 0$	$c_1 = 0.0$	$\gamma_1 = 0$	$c_1 = 0.0$
$\gamma_2 = 1$	$c_2 = 0.589980$	$\gamma_2 = 1$	$c_2 = 0.865164$
$\gamma_3 = 0$	$c_3 = 0.0$	$\gamma_3 = 0$	$c_3 = 0.0$
$\gamma_4 = -0.50$	$c_4 = 0.219469$	$\gamma_4 = -0.50$	$c_4 = -0.147137$
$\gamma_5 = 0$	$c_5 = 0.0$	$\gamma_5 = 0$	$c_5 = 0.0$
$\gamma_6 = 0.455791$	$c_6 = 0.0$	$\gamma_6 = 1.0$	$c_6 = 0.100329$
Height	0.489297	Height	0.375849
Mode	0.0	Mode	±0.543804
Median	0.0	Median	0.0
20% trimmed mean	0.0	20% trimmed mean	0.0
Percentiles		*Percentiles*	
0.01	−2.0747	0.01	−2.1626
0.05	−1.7510	0.05	−1.7149
0.25	−0.6534	0.25	−0.7025
0.75	0.6534	0.75	0.7025
0.95	1.7510	0.95	1.7149
0.99	2.0747	0.99	2.1626

TABLE 3.33

Third- and Fifth-Order Symmetric Uniform-Based Power Method PDFs and Their Associated Cumulants, Coefficients, and Other Indices

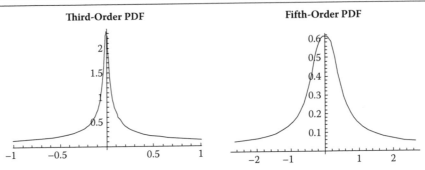

Cumulants	Coefficients	Cumulants	Coefficients
$\gamma_1 = 0$	$c_1 = 0.0$	$\gamma_1 = 0$	$c_1 = 0.0$
$\gamma_2 = 1$	$c_2 = 0.123459$	$\gamma_2 = 1$	$c_2 = 0.469928$
$\gamma_3 = 0$	$c_3 = 0.0$	$\gamma_3 = 0$	$c_3 = 0.0$
$\gamma_4 = 0.50$	$c_4 = 0.450940$	$\gamma_4 = 0.50$	$c_4 = 0.042687$
$\gamma_5 = 0$	$c_5 = 0.0$	$\gamma_5 = 0$	$c_5 = 0.0$
$\gamma_6 = -6.87039$	$c_6 = 0.0$	$\gamma_6 = -6.0$	$c_6 = 0.104451$
Height	2.338233	Height	0.614296
Mode	0.0	Mode	0.0
Median	0.0	Median	0.0
20% trimmed mean	0.0	20% trimmed mean	0.0
Percentiles		*Percentiles*	
0.01	−2.4149	0.01	−2.4782
0.05	−1.9006	0.05	−1.8557
0.25	−0.3998	0.25	−0.4856
0.75	0.3998	0.75	0.4856
0.95	1.9006	0.95	1.8557
0.99	2.4149	0.99	2.4782

TABLE 3.34

Third- and Fifth-Order Symmetric Uniform-Based Power Method PDFs and Their Associated Cumulants, Coefficients, and Other Indices

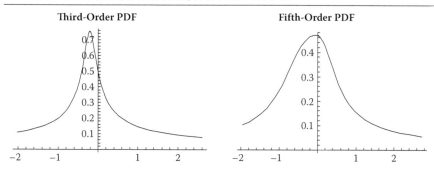

Third-Order PDF		Fifth-Order PDF	
Cumulants	Coefficients	Cumulants	Coefficients
$\gamma_1 = 0$	$c_1 = -0.147113$	$\gamma_1 = 0$	$c_1 = -0.091480$
$\gamma_2 = 1$	$c_2 = 0.406491$	$\gamma_2 = 1$	$c_2 = 0.613360$
$\gamma_3 = 0.50$	$c_3 = 0.147113$	$\gamma_3 = 0.50$	$c_3 = -0.008047$
$\gamma_4 = 0.0$	$c_4 = 0.308216$	$\gamma_4 = 0.0$	$c_4 = 0.068574$
$\gamma_5 = -1.894233$	$c_5 = 0.0$	$\gamma_5 = -1.50$	$c_5 = 0.055293$
$\gamma_6 = -5.280170$	$c_6 = 0.0$	$\gamma_6 = -4.50$	$c_6 = 0.060107$
Height	0.753554	Height	0.470878
Mode	−0.209304	Mode	−0.068921
Median	−0.147113	Median	−0.091480
20% trimmed mean	−0.094152	20% trimmed mean	−0.081479
Percentiles		*Percentiles*	
0.01	−1.9205	0.01	−1.8791
0.05	−1.5908	0.05	−1.5537
0.25	−0.5890	0.25	−0.6714
0.75	0.5154	0.75	0.5386
0.95	2.0116	0.95	1.9845
0.99	2.4741	0.99	2.5678

TABLE 3.35

Third- and Fifth-Order Symmetric Uniform-Based Power Method PDFs and Their Associated Cumulants, Coefficients, and Other Indices

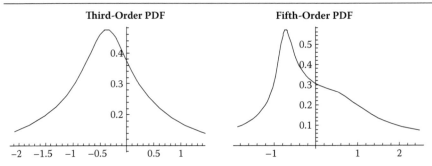

Cumulants	Coefficients	Cumulants	Coefficients
$\gamma_1 = 0$	$c_1 = -0.169269$	$\gamma_1 = 0$	$c_1 = -0.188430$
$\gamma_2 = 1$	$c_2 = 0.662283$	$\gamma_2 = 1$	$c_2 = 0.867465$
$\gamma_3 = 0.50$	$c_3 = 0.169269$	$\gamma_3 = 0.50$	$c_3 = 0.221084$
$\gamma_4 = -0.50$	$c_4 = 0.175828$	$\gamma_4 = -0.50$	$c_4 = -0.109194$
$\gamma_5 = -2.318433$	$c_5 = 0.0$	$\gamma_5 = -2.25$	$c_5 = -0.018141$
$\gamma_6 = -1.423862$	$c_6 = 0.0$	$\gamma_6 = -1.0$	$c_6 = 0.079821$
Height	0.474822	Height	0.571987
Mode	-0.370174	Mode	-0.720336
Median	-0.169269	Median	-0.188430
20% trimmed mean	-0.108332	20% trimmed mean	-0.113072
Percentiles		*Percentiles*	
0.01	-1.6656	0.01	-1.7652
0.05	-1.4563	0.05	-1.4317
0.25	-0.7301	0.25	-0.7520
0.75	0.6455	0.75	0.6864
0.95	1.9405	0.95	1.9151
0.99	2.3025	0.99	2.3612

TABLE 3.36

Third- and Fifth-Order Symmetric Uniform-Based Power Method PDFs and Their Associated Cumulants, Coefficients, and Other Indices

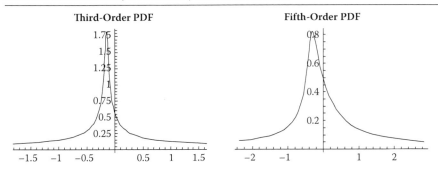

Cumulants	Coefficients	Cumulants	Coefficients
$\gamma_1 = 0$	$c_1 = -0.133794$	$\gamma_1 = 0$	$c_1 = -0.156670$
$\gamma_2 = 1$	$c_2 = 0.172056$	$\gamma_2 = 1$	$c_2 = 0.437453$
$\gamma_3 = 0.50$	$c_3 = 0.133794$	$\gamma_3 = 0.50$	$c_3 = 0.190587$
$\gamma_4 = 0.50$	$c_4 = 0.424007$	$\gamma_4 = 0.50$	$c_4 = -0.102720$
$\gamma_5 = -1.504443$	$c_5 = 0.0$	$\gamma_5 = -1.50$	$c_5 = -0.018843$
$\gamma_6 = -8.644269$	$c_6 = 0.0$	$\gamma_6 = -8.0$	$c_6 = 0.083461$
Height	1.827250	Height	0.825321
Mode	-0.150904	Mode	-0.307225
Median	-0.133794	Median	-0.156670
20% trimmed mean	-0.085628	20% trimmed mean	-0.092454
Percentiles		*Percentiles*	
0.01	-2.1140	0.01	-2.1849
0.05	-1.6831	0.05	-1.6440
0.25	-0.4579	0.25	-0.5105
0.75	0.3910	0.75	0.4619
0.95	2.0657	0.95	2.0344
0.99	2.6175	0.99	2.6569

TABLE 3.37

Third-Order Uniform-Based Power Method PDF Approximations of Two
Multinomial Distributions

Distribution 1	Distribution 2
Number of trials: $n = 1$	Number of trials: $n = 1$
Probabilities:	Probabilities:
$p_1 = .10$ $p_2 = .10$ $p_3 = .18$ $p_4 = .18$ $p_5 = .18$	$p_1 = .10$ $p_2 = .13$ $p_3 = .15$ $p_4 = .18$ $p_5 = .18$
$p_6 = .15$ $p_7 = .11$	$p_6 = .15$ $p_7 = .11$

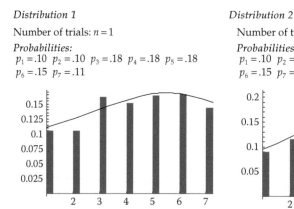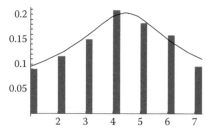

Cumulants	Coefficients	Cumulants	Coefficients
$m = 4.706610$	$c_1 = 0.061873$	$m = 4.648875$	$c_1 = 0.047918$
$s = 1.677016$	$c_2 = 0.941590$	$s = 1.554647$	$c_2 = 0.819068$
$\gamma_3 = -0.155090$	$c_3 = -0.061873$	$\gamma_3 = -0.131893$	$c_3 = -0.047918$
$\gamma_4 = -1.100390$	$c_4 = 0.031429$	$\gamma_4 = -0.912116$	$c_4 = 0.098345$
$\gamma_5 = 0.858754$	$c_5 = 0.0$	$\gamma_5 = 0.679930$	$c_5 = 0.0$
$\gamma_6 = 5.691760$	$c_6 = 0.0$	$\gamma_6 = 3.955880$	$c_6 = 0.0$

Note: Samples of size $N = 500$ were drawn from the two populations.

simultaneously solving two equations of the form $low = m + s \sum_{i=1}^{r} c_i u^{i-1}$ and $upp = m + s \sum_{i=1}^{r} c_i u^{i-1}$ for the power method pdfs' mean (m) and standard deviation (s). For example, if we wanted the power method pdf associated with the first distribution in Table 3.37 (Distribution 1) to begin and end at the lower and upper limits of 1.5 and 7.5, then this would require a mean and standard deviation of $m = 4.70061$ and $s = 2.22935$.

Table 3.38 gives another example of uniform-based approximations to data. This particular data set follows a distribution known as Benford's (1938) law, where the probability of obtaining one of the first nine natural numbers is approximately $p_i = \log_{10}(1 + 1/d_i)$ and where $i = 1, ..., 9$. Benford's law is often used by auditors as a tool to assist in investigating fraud as data associated with some financial transactions; for example, purchases, checking accounts, and so forth, follow this law. Inspection of the graphs in Table 3.38 indicates that the fifth-order polynomial provides a better approximation to the data than the third-order polynomial.

TABLE 3.38

Third- and Fifth-Order Uniform-Based Power Method PDF Approximations and Their Associated Cumulants, Coefficients, and Other Indices

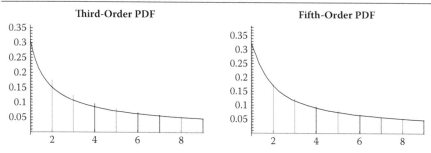

Third-Order PDF		Fifth-Order PDF	
Cumulants	Coefficients	Cumulants	Coefficients
$\hat{\mu} = 3.224428$	$c_1 = -0.3465887$	$\hat{\mu} = 3.377992$	$c_1 = -0.326463$
$\hat{\sigma} = 2.561497$	$c_2 = 0.8530781$	$\hat{\sigma} = 2.522455$	$c_2 = 0.845049$
$\gamma_3 = 0.7978488$	$c_3 = 0.3465887$	$\gamma_3 = 0.7978488$	$c_3 = 0.331413$
$\hat{\gamma}_4 = -0.5067745$	$c_4 = 0.0537341$	$\gamma_4 = -0.5467745$	$c_4 = 0.061175$
$\gamma_5 = -1.504443$	$c_5 = 0.0$	$\hat{\gamma}_5 = -4.024066$	$c_5 = -0.002750$
$\gamma_6 = -8.644269$	$c_6 = 0.0$	$\gamma_6 = -4.0844773$	$c_6 = 0.000156$
Height	0.828255	Height	0.361494
Mode	0.50	Mode	0.50
Median	2.336642	Median	2.554503
20% trimmed mean	2.656245	20% trimmed mean	2.853836
Percentiles		*Percentiles*	
0.00	0.50	0.00	0.50
0.01	0.5123	0.01	0.5271
0.05	0.5663	0.05	0.6340
0.25	1.0207	0.25	1.2312
0.75	4.9843	0.75	5.1241
0.95	8.4217	0.95	8.4560
0.99	9.2768	0.99	9.2840
1.00	9.50	1.00	9.50

Note: The probability of obtaining an integer is $p_i = \log_{10}(1 + 1/d_i)$ for $i = 1, ..., 9$.

3.3 Remediation Techniques

It was demonstrated above that one remedy to the problem of having an invalid power method pdf was to increase the value of kurtosis to a point where the polynomial will produce a valid pdf (e.g., the third-order power method approximation to the beta distribution; Table 3.1). However, it is important to evaluate the appropriateness of any particular adjustment to kurtosis (i.e., $\gamma_4 \rightarrow \hat{\gamma}_4$). A convenient approach for evaluating an adjustment of this type would be to consider the sampling distribution of γ_4 using bootstrapping techniques.

Specifically, Table 3.39 gives estimates of the parameter γ_4 and an associated 95% bootstrap confidence interval for each theoretical distribution where the kurtosis had to be adjusted for the third-order polynomials to produce valid pdfs in Tables 3.1–3.15. The confidence intervals were created by randomly sampling $N = 1000$ data points from each of these distributions and subsequently generating 25,000 bootstrap samples using S-Plus (2007). Inspection of Table 3.39 indicates that the adjustments in kurtosis associated with the beta distributions are not appropriate because the values of $\hat{\gamma}_4$ used all lie above the upper limit of the 95% confidence interval. On the other hand, the adjustments to kurtosis associated with the chi-square, Weibull, and F distributions would be considered appropriate because the values of $\hat{\gamma}_4$ that were used all lie well within the 95% confidence interval. For example, in terms of the $F_{3,100}$ distribution, approximately 20% of the bootstrap samples of kurtosis were greater than the value of $\hat{\gamma}_4 = 4.905548$ that was used.

In the context of fifth-order polynomials, the simplest remedy to the problem of having an invalid pdf is to increase the sixth cumulant $\gamma_6 \rightarrow \tilde{\gamma}_6$ and

TABLE 3.39

Estimates and 95% Bootstrap Confidence Intervals on Samples of Data ($N = 1000$) Drawn from the Specified Populations

Population	γ_4	$\hat{\gamma}_4$	Estimate (95% Bootstrap CI)
Beta(2,4)	−0.375	0.345	−0.340 (−0.533, −0.100)*
Beta(4,4)	−0.545455	0.004545	−0.513 (−0.638, −0.346)*
Beta(5,4)	−0.477273	0.032727	−0.533 (−0.666, −0.375)*
χ_3^2	4.0	4.21	3.662 (2.260, 5.752)
χ_6^2	2.0	2.09	1.962 (0.975, 3.530)
χ_8^2	1.5	1.57	1.574 (0.968, 2.264)
Weibull(2,5)	0.245089	0.625089	0.185 (−0.299, 1.346)
Weibull(6,5)	0.035455	0.225455	0.085 (−0.155, 0.438)
Weibull(10,5)	0.570166	0.640166	0.568 (0.043, 1.997)
$F_{3,100}$	4.885548	4.905548	5.271 (2.659, 9.444)

Note: The value γ_4 is the population kurtosis, and $\hat{\gamma}_4$ was the required value of kurtosis used to ensure a valid third-order power method pdf.

* Indicates that the estimate of γ_4 is not contained within the 95% Bootstrap cI.

thereby preserve the values of the lower cumulants. Some examples are provided in Tables 3.7–3.9, where the values of γ_6 had to be increased by 1.0 to ensure valid power method pdfs associated with the Weibull distributions. Because γ_6 has high variance, such small alterations to this cumulant will usually have little impact on a power method approximation.

There are occasions when a simple exogenous adjustment to γ_6 will not remedy the problem of an invalid fifth-order power method pdf. When this occurs, there are some additional steps that can be taken to potentially produce a valid pdf. Specifically, it is first necessary to consider the values of skew (γ_3) and kurtosis (γ_4) in the context of a third-order polynomial. As indicated above, attempt to increase the value of kurtosis $\gamma_4 \to \hat{\gamma}_4$ to the point where a valid third-order pdf is obtained. If a valid power method is produced, then use the solved coefficients $(c_1, ..., c_4)$, associated with the third-order pdf to determine the values of the fifth and sixth cumulants by setting $c_5 = c_6 = 0$ in the equations for γ_5 and γ_6 associated with the fifth-order system. Subsequently using the fifth-order system, the user can alter the values of γ_5 and γ_6 until a potential good fit to a theoretical pdf or data set is made. Tables 3.40 and 3.41 provide examples of normal- and logistic-based power

TABLE 3.40

Third- and Fifth-Order Normal-Based Power Method Remedial Approximations of a Lognormal Data Set ($N = 250$)

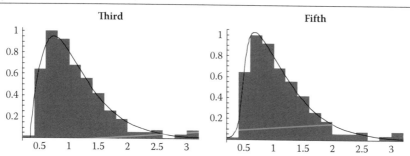

	Cumulants			Coefficients	
Data	Third	Fifth		Third	Fifth
$\mu = 1.112432$	$\mu = 1.112432$	$\mu = 1.112432$		$c_1 = -0.224448$	$c_1 = -0.240798$
$\sigma^2 = 0.302226$	$\sigma^2 = 0.302226$	$\sigma^2 = 0.302226$		$c_2 = 0.890078$	$c_2 = 0.897507$
$\gamma_3 = 1.452875$	$\gamma_3 = 1.452875$	$\gamma_3 = 1.452875$		$c_3 = 0.224448$	$c_3 = 0.258827$
$\gamma_4 = 2.52083$	$\hat{\gamma}_4 = 3.32$	$\hat{\gamma}_4 = 3.32$		$c_4 = 0.019021$	$c_4 = 0.007140$
(1.377, 3.832)					
$\gamma_5 = 2.28162$	$\hat{\gamma}_5 = 10.5729$	$\hat{\gamma}_5 = 10.5$		$c_5 = 0.0$	$c_5 = -0.006010$
$\gamma_6 = -14.2288$	$\hat{\gamma}_6 = 43.5467$	$\hat{\gamma}_6 = 50$		$c_6 = 0.0$	$c_6 = 0.001829$

Note: The value of γ_4 has an associated 95% bootstrap CI enclosed in parentheses.

TABLE 3.41

Third- and Fifth-Order Logistic-Based Power Method Remedial Approximations of a Lognormal Data Set ($N = 250$)

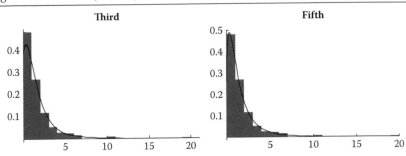

Data	Cumulants		Coefficients	
	Third	Fifth	Third	Fifth
$\mu = 1.623663$	$\mu = 1.623663$	$\mu = 1.623663$	$c_1 = -0.271725$	$c_1 = -0.277632$
$\sigma^2 = 4.048619$	$\sigma^2 = 4.048619$	$\sigma^2 = 4.048619$	$c_2 = 0.686430$	$c_2 = 0.645446$
$\gamma_3 = 4.237661$	$\gamma_3 = 4.237661$	$\gamma_3 = 4.237661$	$c_3 = 0.271725$	$c_3 = 0.286539$
$\gamma_4 = 28.9103$	$\hat{\gamma}_4 = 45$	$\hat{\gamma}_4 = 45$	$c_4 = 0.039791$	$c_4 = 0.048319$
(7.59, 64.93)				
$\gamma_5 = 227.600$	$\hat{\gamma}_5 = 922.264$	$\hat{\gamma}_5 = 740$	$c_5 = 0.0$	$c_5 = -0.002121$
$\gamma_6 = 1848.08$	$\hat{\gamma}_6 = 32334.8$	$\hat{\gamma}_6 = 30000$	$c_6 = 0.0$	$c_6 = 0.000050$

Note: The value of γ_4 has an associated 95% bootstrap CI enclosed in parentheses.

method approximations to lognormal data sets, respectively. It is noted that logistic-based polynomials were used on the data associated with Table 3.41 because normal-based polynomials could not produce valid power method pdfs for the given values of skew and kurtosis.

3.4 Monte Carlo Simulation

In this section the results of a Monte Carlo simulation are provided to evaluate the methodology presented in the previous chapter in terms of accuracy and error. More specifically, Table 3.42 gives the standardized cumulants for six power method pdfs based on uniform (W_1, W_4), normal (W_2, W_5), and logistic distributions (W_3, W_6).

These distributions are the approximations to the multinomial, chi-square, and t_3 distributions in Tables 3.37 (multinomial), 3.4 and 3.5 (chi-square),

TABLE 3.42

Population Cumulants for the Distributions in the Monte Carlo Simulation.
The Distributions Are (a) $W_{1,4}$ = Uniform, (b) $W_{2,5}$ = Normal, and (c) $W_{3,6}$ = Logistic.

	γ_4	γ_2	γ_3	γ_4	γ_5	γ_6
$p(W_1)$	0.0	1.0	−0.155090	−1.10039	0.858754	5.69176
$p(W_2)$	0.0	1.0	1.63299	4.0	13.0639	53.3333
$p(W_3)$	0.0	1.0	−1.73390	21.9084	−166.334	12087.3
$p(W_4)$	0.0	1.0	−0.131893	−0.912116	0.679930	3.95588
$p(W_5)$	0.0	1.0	1.15470	2.0	4.61880	13.3333
$p(W_6)$	0.0	1.0	0.168484	21.1908	56.2067	13841.8

and 3.30 (t_3), respectively. Table 3.43 gives the specified correlation matrix between the six variables. Using the methodology described in Section 2.6, Tables 3.44 and 3.45 give the computed intermediate correlations and the associated Cholesky factorization.

The simulation was conducted using an algorithm coded in Fortran employing the use of subroutines UN1 and NORMB1 from RANGEN (Blair, 1987) to generate pseudorandom uniform and standard normal deviates. Independent samples sizes of $N = 10$, 25, 50, 100, 250, 25,000, and 100,000 were generated for simulating the specified cumulants and correlations. The averaging procedure used to compute the estimates of the cumulants and correlations was based on 25,000 replications and thus $N \times 25000$ random deviates.

Table 3.46 gives the overall average estimates of the correlation coefficients from the simulation. These estimates were all in close agreement with the specified parameters, even for sample sizes as small as $N = 10$. As indicated in Table 3.46, across all sample sizes the estimates were all equal to the population parameters when rounding was to three decimal places. In terms of error, presented in Tables 3.47–3.50

TABLE 3.43

Specified Correlations $\rho_{p(W_i)p(W_j)}$ between the Power Method Distributions in the Monte Carlo Simulation

	$p(W_1)$	$p(W_2)$	$p(W_3)$	$p(W_4)$	$p(W_5)$	$p(W_6)$
$p(W_1)$	1					
$p(W_2)$	0.20	1				
$p(W_3)$	0.20	0.40	1			
$p(W_4)$	0.20	0.40	0.50	1		
$p(W_5)$	0.20	0.40	0.50	0.60	1	
$p(W_6)$	0.20	0.40	0.50	0.60	0.70	1

Note: The polynomials are based on $p(W_{1,4}$:= Uniform), $p(W_{2,5}$:= Normal), and $p(W_{3,6}$:= Logistic) distributions.

TABLE 3.44

Intermediate Correlation Matrix Associated with Table 3.43

	Z_1	Z_2	Z_3	Z_4	Z_5	Z_6
Z_1	1					
Z_2	0.219945	1				
Z_3	0.218990	0.475633	1			
Z_4	0.206348	0.437989	0.548427	1		
Z_5	0.212271	0.425785	0.572471	0.635148	1	
Z_6	0.216396	0.450663	0.549758	0.656256	0.761209	1

TABLE 3.45

Cholesky Decomposition on the Intermediate Correlation Matrix in Table 3.44

$a_{11}=1$	$a_{12}=0.219945$	$a_{13}=0.218990$	$a_{14}=0.206348$	$a_{15}=0.212271$	$a_{16}=0.216396$
0	$a_{22}=0.975512$	$a_{23}=0.438198$	$a_{24}=0.402459$	$a_{25}=0.388613$	$a_{26}=0.413186$
0	0	$a_{33}=0.871795$	$a_{34}=0.374953$	$a_{35}=0.408005$	$a_{36}=0.368564$
0	0	0	$a_{44}=0.809233$	$a_{45}=0.348432$	$a_{46}=0.379518$
0	0	0	0	$a_{55}=0.718365$	$a_{56}=0.378767$
0	0	0	0	0	$a_{66}=0.599260$

TABLE 3.46

Average Correlations (Rounded to Three Decimal Points) Associated with the Specified Correlations in Table 3.42 for $N = 10, 25, 50, 100, 250,$ and $1,000$

	$p(W_1)$	$p(W_2)$	$p(W_3)$	$p(W_4)$	$p(W_5)$	$p(W_6)$
$p(W_1)$	1					
$p(W_2)$	0.200	1				
$p(W_3)$	0.200	0.400	1			
$p(W_4)$	0.200	0.400	0.500	1		
$p(W_5)$	0.200	0.400	0.500	0.600	1	
$p(W_6)$	0.200	0.400	0.500	0.600	0.700	1

TABLE 3.47

Mean Absolute Deviations Associated with the Average Correlations in Table 3.46 with $N = 10$

	$p(W_1)$	$p(W_2)$	$p(W_3)$	$p(W_4)$	$p(W_5)$	$p(W_6)$
$p(W_1)$	1					
$p(W_2)$	0.244	1				
$p(W_3)$	0.243	0.268	1			
$p(W_4)$	0.251	0.256	0.285	1		
$p(W_5)$	0.247	0.295	0.293	0.269	1	
$p(W_6)$	0.197	0.287	0.349	0.299	0.379	1

TABLE 3.48

Mean Absolute Deviations Associated with the Average Correlations in Table 3.46 with $N = 100$

	$p(W_1)$	$p(W_2)$	$p(W_3)$	$p(W_4)$	$p(W_5)$	$p(W_6)$
$p(W_1)$	1					
$p(W_2)$	0.079	1				
$p(W_3)$	0.083	0.093	1			
$p(W_4)$	0.080	0.084	0.097	1		
$p(W_5)$	0.080	0.104	0.103	0.086	1	
$p(W_6)$	0.082	0.105	0.138	0.101	0.141	1

TABLE 3.49

Mean Absolute Deviations Associated with the Average Correlations in Table 3.46 with $N = 250$

	$p(W_1)$	$p(W_2)$	$p(W_3)$	$p(W_4)$	$p(W_5)$	$p(W_6)$
$p(W_1)$	1					
$p(W_2)$	0.050	1				
$p(W_3)$	0.052	0.061	1			
$p(W_4)$	0.050	0.053	0.062	1		
$p(W_5)$	0.050	0.067	0.066	0.055	1	
$p(W_6)$	0.052	0.068	0.092	0.064	0.091	1

TABLE 3.50

Mean Absolute Deviations Associated with the Average Correlations in Table 3.46 with $N = 1000$

	$p(W_1)$	$p(W_2)$	$p(W_3)$	$p(W_4)$	$p(W_5)$	$p(W_6)$
$p(W_1)$	1					
$p(W_2)$	0.025	1				
$p(W_3)$	0.026	0.031	1			
$p(W_4)$	0.025	0.027	0.031	1		
$p(W_5)$	0.025	0.033	0.033	0.027	1	
$p(W_6)$	0.026	0.035	0.049	0.032	0.047	1

TABLE 3.51

Empirical Estimates of the Population Cumulants in Table 3.42

	$\hat{\gamma}_1$	$\hat{\gamma}_2$	$\hat{\gamma}_3$	$\hat{\gamma}_4$	$\hat{\gamma}_5$	$\hat{\gamma}_6$
$p(W_1)$	0.000	0.999	−0.155	−1.102	0.857	5.689
$p(W_2)$	0.000	1.000	1.633	4.005	13.111	53.676
$p(W_3)$	0.000	0.998	−1.733	21.238	−218.181	8360.487
$p(W_4)$	0.000	0.999	−0.132	−0.914	0.678	3.951
$p(W_5)$	0.000	0.999	1.153	1.997	4.611	13.296
$p(W_6)$	0.000	0.999	0.156	20.040	32.765	6928.011

Note: The estimates are based on samples of size $N = 10$.

are the mean absolute deviations (MDs) for the correlation coefficients, where $MD = \Sigma |\hat{\rho}_{ij} - \rho_{ij}|/25000$. Inspection of these tables indicates that larger (absolute) values of the population correlations are associated with a larger amount of error. For example, for $N = 100$ the values of MD associated with $\rho = 0.20$ are approximately 0.08, whereas the value of MD for $\rho = 0.70$ is 0.141. Further, the magnitude of the MD is also contingent on the shape of the distributions. For example, distributions with larger values of the higher-order cumulants (e.g., Distributions 3 and 6) had larger MD values.

Presented in Tables 3.51–3.58 are the estimated cumulants for the six distributions for the various sample sizes. The rate at which the estimates converged to their population parameter was contingent on shape of the distribution. Essentially, higher-order cumulants ($\gamma_{4,5,6}$) with large values require larger sample sizes for the estimates to obtain close proximity to their associated parameters, as opposed to cumulants of lower order ($\gamma_{1,2,3}$) or cumulants of higher order ($\gamma_{4,5,6}$) but with smaller values. For example, the cumulants $\gamma_{1,...,6}$ for the uniform-based polynomials were all in close proximity with their respective parameter, even for sample sizes as small as $N = 10$. On the other hand, the logistic-based polynomials required the largest sample sizes to

TABLE 3.52

Empirical Estimates of the Population Cumulants in Table 3.42

	$\hat{\gamma}_1$	$\hat{\gamma}_2$	$\hat{\gamma}_3$	$\hat{\gamma}_4$	$\hat{\gamma}_5$	$\hat{\gamma}_6$
$p(W_1)$	0.000	1.000	−0.155	−1.101	0.857	5.692
$p(W_2)$	0.000	1.000	1.631	3.990	12.996	52.734
$p(W_3)$	0.000	0.999	−1.745	21.105	−216.233	7487.681
$p(W_4)$	0.000	1.000	−0.132	−0.912	0.678	3.956
$p(W_5)$	0.000	1.000	1.153	1.990	4.556	12.885
$p(W_6)$	0.000	0.999	0.175	20.370	24.508	9088.067

Note: The estimates are based on samples of size $N = 25$.

TABLE 3.53

Empirical Estimates of the Population Cumulants in Table 3.42

	$\hat{\gamma}_1$	$\hat{\gamma}_2$	$\hat{\gamma}_3$	$\hat{\gamma}_4$	$\hat{\gamma}_5$	$\hat{\gamma}_6$
$p(W_1)$	0.000	1.000	−0.155	−1.100	0.857	5.693
$p(W_2)$	0.000	1.000	1.632	3.996	13.040	52.997
$p(W_3)$	0.000	0.999	−1.737	20.884	−200.498	6852.837
$p(W_4)$	0.000	1.000	−0.132	−0.912	0.678	3.957
$p(W_5)$	0.000	1.000	1.154	1.992	4.568	12.982
$p(W_6)$	0.000	1.000	0.172	20.576	32.838	9056.522

Note: The estimates are based on samples of size $N = 50$.

TABLE 3.54

Empirical Estimates of the Population Cumulants in Table 3.42

	$\hat{\gamma}_1$	$\hat{\gamma}_2$	$\hat{\gamma}_3$	$\hat{\gamma}_4$	$\hat{\gamma}_5$	$\hat{\gamma}_6$
$p(W_1)$	0.000	1.000	−0.155	−1.100	0.857	5.692
$p(W_2)$	0.000	1.000	1.632	3.993	13.004	52.686
$p(W_3)$	0.000	1.000	−1.748	21.625	−206.494	8327.573
$p(W_4)$	0.000	1.000	−0.132	−0.912	0.678	3.957
$p(W_5)$	0.000	1.000	1.154	1.995	4.584	13.071
$p(W_6)$	0.000	1.000	0.1750	20.465	45.043	8505.382

Note: The estimates are based on samples of size $N = 100$.

TABLE 3.55

Empirical Estimates of the Population Cumulants in Table 3.42

	$\hat{\gamma}_1$	$\hat{\gamma}_2$	$\hat{\gamma}_3$	$\hat{\gamma}_4$	$\hat{\gamma}_5$	$\hat{\gamma}_6$
$p(W_1)$	0.000	1.000	−0.155	−1.100	0.856	5.693
$p(W_2)$	0.000	1.000	1.632	3.996	13.025	52.885
$p(W_3)$	0.000	1.000	−1.750	21.852	−209.260	8713.399
$p(W_4)$	0.000	1.000	−0.132	−0.912	0.679	3.956
$p(W_5)$	0.000	1.000	1.155	1.999	4.613	13.348
$p(W_6)$	0.000	1.000	0.175	20.552	54.467	8728.382

Note: The estimates are based on samples of size $N = 250$.

TABLE 3.56

Empirical Estimates of the Population Cumulants in Table 3.42

	$\hat{\gamma}_1$	$\hat{\gamma}_2$	$\hat{\gamma}_3$	$\hat{\gamma}_4$	$\hat{\gamma}_5$	$\hat{\gamma}_6$
$p(W_1)$	0.000	1.000	−0.155	−1.101	0.858	5.690
$p(W_2)$	0.000	0.999	1.630	3.977	12.872	51.442
$p(W_3)$	0.000	1.001	−1.747	21.327	−187.082	7076.181
$p(W_4)$	0.000	1.000	−0.131	−0.912	0.681	3.956
$p(W_5)$	0.000	1.001	1.157	2.007	4.639	13.375
$p(W_6)$	0.000	1.000	0.169	21.186	53.138	11788.130

Note: The estimates are based on samples of size $N = 1000$.

TABLE 3.57

Empirical Estimates of the Population Cumulants in Table 3.42

	$\hat{\gamma}_1$	$\hat{\gamma}_2$	$\hat{\gamma}_3$	$\hat{\gamma}_4$	$\hat{\gamma}_5$	$\hat{\gamma}_6$
$p(W_1)$	0.000	1.000	−0.155	−1.100	0.857	5.692
$p(W_2)$	0.000	1.000	1.633	3.998	13.042	53.006
$p(W_3)$	0.000	1.000	−1.738	21.377	−184.957	7666.786
$p(W_4)$	0.000	1.000	−0.132	−0.912	0.679	3.956
$p(W_5)$	0.000	1.000	1.155	1.999	4.614	13.286
$p(W_6)$	0.000	1.000	0.168	20.568	45.407	8817.380

Note: The estimates are based on samples of size $N = 25000$.

TABLE 3.58

Empirical Estimates of the Population Cumulants in Table 3.42

	$\hat{\gamma}_1$	$\hat{\gamma}_2$	$\hat{\gamma}_3$	$\hat{\gamma}_4$	$\hat{\gamma}_5$	$\hat{\gamma}_6$
$p(W_1)$	0.000	1.000	−0.155	−1.100	0.857	5.692
$p(W_2)$	0.000	1.000	1.633	4.000	13.042	52.970
$p(W_3)$	0.000	1.000	−1.734	21.367	−185.450	7555.936
$p(W_4)$	0.000	1.000	−0.132	−0.912	0.679	3.956
$p(W_5)$	0.000	1.000	1.155	2.000	4.618	13.327
$p(W_6)$	0.000	1.000	0.168	20.649	46.059	9101.455

Note: The estimates are based on samples of size $N = 100000$.

TABLE 3.59

Mean Absolute Deviations (md) for the Cumulants Associated with Variable $p(W_1)$ in Table 3.42

N	md$\{\gamma_1\}$	md$\{\gamma_2\}$	md$\{\gamma_3\}$	md$\{\gamma_4\}$	md$\{\gamma_5\}$	md$\{\gamma_6\}$
10	0.254	0.240	0.527	0.710	1.402	2.166
25	0.160	0.152	0.334	0.448	0.892	1.370
50	0.113	0.107	0.237	0.317	0.632	0.968
100	0.080	0.076	0.167	0.224	0.447	0.685
250	0.050	0.048	0.106	0.142	0.283	0.433
1,000	0.025	0.024	0.053	0.071	0.141	0.218
25,000	0.005	0.005	0.011	0.014	0.028	0.043
100,000	0.002	0.002	0.005	0.007	0.014	0.020

obtain estimates that were in close agreement with the parameters. It should also be pointed out that the sixth cumulant, γ_6, associated with the logistic-based polynomials failed to provide estimates that were close to their respective parameter. These observations are consistent with the results reported in Tables 3.59–3.64 for the mean absolute deviations. Specifically, the uniform polynomials produced the smallest values of MD (MD $= \Sigma |\hat{\gamma}_i - \gamma_i|/25000$), whereas the logistic polynomials had the largest values. Further, inspection of Tables 3.61 and 3.64 indicates that the logistic-based polynomials $p(W_3)$ and $p(W_6)$ required sample sizes as large as 100,000 to yield substantial reductions in the MD for the sixth cumulant γ_6.

TABLE 3.60

Mean Absolute Deviations (md) for the Cumulants Associated with Variable $p(W_2)$ in Table 3.42

N	md$\{\gamma_1\}$	md$\{\gamma_2\}$	md$\{\gamma_3\}$	md$\{\gamma_4\}$	md$\{\gamma_5\}$	md$\{\gamma_6\}$
10	0.251	0.529	2.028	8.756	44.037	253.348
25	0.159	0.362	1.503	7.016	37.484	224.439
50	0.113	0.264	1.159	5.735	32.183	200.199
100	0.080	0.191	0.871	4.545	26.803	173.883
250	0.051	0.122	0.582	3.230	20.296	139.610
1,000	0.025	0.062	0.303	1.793	12.245	91.469
25,000	0.005	0.012	0.062	0.393	3.034	26.472
100,000	0.002	0.006	0.031	0.198	1.564	14.205

TABLE 3.61

Mean Absolute Deviations (md) for the Cumulants Associated with Variable $p(W_3)$ in Table 3.42

N	md$\{\gamma_1\}$	md$\{\gamma_2\}$	md$\{\gamma_3\}$	md$\{\gamma_4\}$	md$\{\gamma_5\}$	md$\{\gamma_6\}$
10	0.242	0.671	3.556	40.115	511.059	20,559.260
25	0.156	0.496	3.042	36.939	482.198	19,405.960
50	0.112	0.385	2.640	34.040	454.152	18,498.610
100	0.079	0.296	2.270	31.815	466.045	19,625.500
250	0.050	0.203	1.794	27.915	445.300	19,418.050
1,000	0.025	0.111	1.177	21.050	360.817	16,566.840
25,000	0.005	0.025	0.355	9.249	217.728	12,794.130
100,000	0.002	0.012	0.197	5.811	156.916	10,347.620

TABLE 3.62

Mean Absolute Deviations (md) for the Cumulants Associated with Variable $p(W_4)$ in Table 3.42

N	md$\{\gamma_1\}$	md$\{\gamma_2\}$	md$\{\gamma_3\}$	md$\{\gamma_4\}$	md$\{\gamma_5\}$	md$\{\gamma_6\}$
10	0.253	0.264	0.583	0.856	1.730	2.886
25	0.160	0.167	0.371	0.542	1.110	1.828
50	0.113	0.118	0.263	0.383	0.787	1.292
100	0.080	0.083	0.186	0.271	0.557	0.914
250	0.051	0.053	0.118	0.171	0.353	0.578
1,000	0.025	0.026	0.059	0.086	0.177	0.289
25,000	0.005	0.005	0.012	0.017	0.035	0.058
100,000	0.002	0.002	0.006	0.009	0.017	0.029

TABLE 3.63

Mean Absolute Deviations (md) for the Cumulants Associated with Variable $p(W_5)$ in Table 3.42

N	md$\{\gamma_1\}$	md$\{\gamma_2\}$	md$\{\gamma_3\}$	md$\{\gamma_4\}$	md$\{\gamma_5\}$	md$\{\gamma_6\}$
10	0.251	0.448	1.518	5.521	23.363	110.678
25	0.159	0.301	1.093	4.303	19.339	95.790
50	0.113	0.219	0.829	3.446	16.243	83.704
100	0.080	0.157	0.615	2.684	13.263	71.308
250	0.051	0.100	0.405	1.865	9.782	55.745
1,000	0.025	0.050	0.207	1.006	5.670	34.978
25,000	0.005	0.010	0.042	0.217	1.351	9.527
100,000	0.002	0.005	0.002	0.108	0.685	5.001

TABLE 3.64

Mean Absolute Deviations (md) for the Cumulants Associated with Variable $p(W_6)$ in Table 3.42

N	md$\{\gamma_1\}$	md$\{\gamma_2\}$	md$\{\gamma_3\}$	md$\{\gamma_4\}$	md$\{\gamma_5\}$	md$\{\gamma_6\}$
10	0.242	0.654	2.709	38.403	374.904	21,078.990
25	0.156	0.478	2.379	35.819	399.522	22,923.540
50	0.111	0.371	2.123	33.522	397.140	22,635.200
100	0.078	0.283	1.853	30.744	380.301	21,747.480
250	0.050	0.194	1.510	27.146	369.414	21,354.680
1,000	0.025	0.106	1.051	22.184	387.070	23,155.250
25,000	0.005	0.024	0.359	10.011	238.688	15,526.280
100,000	0.002	0.012	0.206	6.598	189.688	13,120.590

3.5 Some Further Considerations

Consider Equation (2.7) with $r = 6$ expressed in the following two algebraically equivalent forms:

$$p(W) = c_1 + c_2 W + c_3 W^2 + c_4 W^3 + c_5 W^4 + c_6 W^5 \tag{3.6}$$

$$p(W) = c_1 + W(c_2 + W(c_3 + W(c_4 + W(c_5 + c_6 W)))) \tag{3.7}$$

If an algorithm was used to generate $p(W)$ in the notation of Equation 3.7 instead of Equation 3.6, then the runtime of a Monte Carlo or simulation study is substantially reduced. To demonstrate, using a Pentium-based PC, it took approximately 25 seconds of computer time to draw 100,000 random samples of size $N = 500$ from a distribution with cumulants associated with a $\chi^2_{df=2}$ distribution using Equation 3.7. On the other hand, using Equation 3.6, the sample size had to be reduced to $N = 100$ to obtain 100,000 draws within the same 25-second time period. Thus, a substantial gain in efficiency can be realized when an algorithm based on Equation 3.7 is used in lieu of Equation 3.6—particularly when the number of variables in the simulation study is large.

Another topic worth discussing is the manner in which two polynomials $p(W_1)$ and $p(W_2)$ are generated in the context of variance reduction. More specifically, a method that can be used to improve the efficiency of the estimate of $(p(W_1) + p(W_2))/2$ is inducing a negative correlation on $p(W_1)$ and $p(W_2)$. To demonstrate, since $p(W_1)$ and $p(W_2)$ are identically distributed with zero means and unit variances, we have

$$\text{Var}[(p(W_1) + p(W_2))/2] = 1/2 + \text{Corr}[p(W_1), p(W_2)]/2 \tag{3.8}$$

By inspection of Equation 3.8 it would be advantageous if $p(W_1)$ and $p(W_2)$ were negatively correlated. To induce a negative correlation on $p(W_1)$ and $p(W_2)$, it is only necessary to simultaneously reverse the signs of the coefficients with even subscripts in $p(W_2)$ as

$$p(W_1) = f_1(c_1, c_2, c_3, c_4, c_5, c_6, Z_1) \qquad (3.9)$$

$$p(W_2) = f_2(c_1, -c_2, c_3, -c_4, c_5, -c_6, Z_1) \qquad (3.10)$$

where Z_1 is standard normal in Equations 3.9 and Equation 3.10. Because $p(W_i)$ both have zero means and unit variances, the correlation between the two polynomials can be expressed as

$$\rho_{p(W_1)p(W_2)} = E[p(W_1)p(W_2)] \qquad (3.11)$$

Expanding Equation 3.11 and taking expectations using the moments from the standard normal distribution yields

$$\rho_{p(W_1)p(W_2)} = c_1^2 - c_2^2 + 2c_1(c_3 + 3c_5) - 6c_2(c_4 + 5c_6)$$

$$+ 3\left(c_3^2 + 10c_3c_5 - 5\left(c_4^2 - 7c_5^2 + 14c_4c_6 + 63c_6^2\right)\right) \qquad (3.12)$$

Inspection of Equation 3.12 reveals that the correlation between $p(W_1)$ and $p(W_2)$ can be determined by evaluating Equation 3.12 using the coefficients from Equation 3.9. For example, evaluating Equation 3.12 using the coefficients that are associated with the standardized cumulants of an exponential distribution (see Table 4.20, Panel B) gives $\rho_{p(W_1)p(W_2)} \cong -0.647$.

This method of inducing a negative correlation between $p(W_1)$ and $p(W_2)$ is analogous to the method used on distributions generated by the inverse transform method. More specifically, consider generating Y_1 and Y_2 from the single-parameter exponential family with distribution function G and with an inverse distribution function denoted as G^{-1}. Let $Y_1 = G^{-1}(V)$ and $Y_2 = G^{-1}(1-V)$, where $V \sim U(0,1)$. Define the parameters for the first and second moments as θ and θ^2. From the definition of the product moment of correlation we have

$$E[Y_1Y_2] = \theta^2 \int_0^1 \ln v \ln(1-v)\,dv = \theta^2(2 - \pi^2/6)$$

As such, the correlation between Y_1 and Y_2 is

$$\rho_{Y_1Y_2} \cong -0.645 \qquad (3.13)$$

Thus, the approximation for the exponential distribution given by Equation 3.12, that is, $\rho_{p(W_1)p(W_2)} \cong -0.647$, is very close to the exact result given in Equation 3.13.

TABLE 3.65

Confidence Intervals (CIs) on the Estimate of $((p(W_1)+p(W_2))/2$ with and without a Negative Correlation Induced

Corr[$p(W_1)$, $p(W_2)$]	Sample Size	95% CI
0.000	$n = 10$	[4.552, 5.448]
−0.647		[4.715, 5.252]
0.000	$n = 26$	[4.726, 5.273]
−0.647		[4.841, 5.158]

Note: $p(W_1)$ and $p(W_2)$ are approximate exponential distributions with population means of 5. The CIs are based on 50,000 sample estimates.

Presented in Table 3.65 are confidence intervals from a simulation (Headrick, 2004) that demonstrate the advantage of inducing a negative correlation on $p(W_1)$ and $p(W_2)$. By inspection of Table 3.65, when $p(W_1)$ and $p(W_2)$ are uncorrelated it takes more than 2.5 times the sample size to obtain a confidence interval that has approximately the same width as the data with an induced negative correlation. Thus, whenever possible, it is advantageous to induce a negative correlation to improve the computational and statistical efficiency of a Monte Carlo study. This technique can also be applied in the context of uniform- and logistic-based polynomials.

4

Simulating More Elaborate Correlation Structures

4.1 Introduction

The primary focus of Section 2.6 was on simulating nonnormal distributions with a specified correlation matrix. In this chapter we extend this focus for the purpose of simulating more elaborate correlation structures. Basically, methods are described in the context of fifth-order power method polynomials for simulating systems of linear statistical models, intraclass correlation coefficients (ICCs), and variate rank correlations.

Specifically, and in terms of systems of linear models, a procedure is described that allows for the simultaneous control of the correlated nonnormal (1) error or stochastic disturbance distributions, (2) independent variables, and (3) dependent and independent variables for each model or equation throughout a system. In terms of ICCs, a procedure is developed for simulating correlated nonnormal distributions with specified (1) standardized cumulants, (2) Pearson intercorrelations, and (3) ICCs. The structural model of concern here is a two-factor design with either fixed or random effects. Finally, in terms of variate rank correlations, a method is presented for simulating controlled correlation structures between nonnormal (1) variates, (2) ranks, and (3) variates with ranks in the context of the power method. The correlation structure between variate values and their associated rank order is also provided for power method transformations. As such, an index of the potential loss of information is provided when ranks are used in place of variate values.

Numerical examples are provided to demonstrate each of the procedures presented. The results from Monte Carlo simulations are also provided to confirm that these procedures generate their associated specified cumulants and correlations.

4.2 Simulating Systems of Linear Statistical Models

In this section, the focus is on the topic of simulating systems of linear statistical models where several models may be considered jointly within a system (e.g., Headrick & Beasley, 2004). Let us first consider a system of T statistical models or equations as

$$\mathbf{y}_p = \mathbf{x}_p \boldsymbol{\beta}_p + \boldsymbol{\sigma}_p \boldsymbol{\varepsilon}_p \tag{4.1}$$

where $p = 1, \ldots, T$, $\mathbf{y}_p$ and $\boldsymbol{\varepsilon}_p$ have dimension $(N \times 1)$, $\mathbf{x}_p$ is $(N \times k_p)$, $\boldsymbol{\beta}_p$ is $(k_p \times 1)$, and $\boldsymbol{\sigma}_p$ is a real positive scalar. The p equations can be expressed as a linear system as

$$\mathbf{y} = \mathbf{x} \boldsymbol{\beta} + \boldsymbol{\sigma} \boldsymbol{\varepsilon} \tag{4.2}$$

where $\mathbf{y}$ and $\boldsymbol{\varepsilon}$ have dimension $(TN \times 1)$, $\mathbf{x}$ is $(TN \times K)$, $\boldsymbol{\beta}$ is $(K \times 1)$, $K = \sum_{p=1}^{T} k_p$, and $\boldsymbol{\sigma} = (\boldsymbol{\sigma}_1, \ldots, \boldsymbol{\sigma}_T)$ represents T scalars associated with each of the T equations. The error terms $\boldsymbol{\varepsilon}$ associated with Equation 4.2 are assumed to have zero means and unit variances.

If the error terms in Equation 4.2 are correlated (e.g., $\boldsymbol{\varepsilon}_p$ is correlated with $\boldsymbol{\varepsilon}_q$), then a gain in efficiency can be realized by using the method of generalized least squares (GLS) as opposed to ordinary least squares (OLS) (Judge, Hill, Griffiths, Lutkepohl, & Lee, 1988). This approach of joint estimation is perhaps better known as "seemingly unrelated regression equation estimation" (Zellner, 1962). The stronger (weaker) the correlations are between the error terms (independent variables) in Equation 4.2, then the greater the efficiency GLS has relative to OLS (Dwivedi & Srivastava, 1978). Thus, it may be desirable to study the relative Type I error and power properties of the GLS and OLS estimators under nonnormal conditions. Such an investigation would usually be carried out using Monte Carlo techniques. To determine the advantages of GLS relative to OLS, a variety of nonnormal distributions with varying degrees of correlation between the error terms would usually be included in the study.

There are many methods and applications of linear models that involve a set of statistical equations. Some examples include confirmatory factor analysis, generalized linear models, hierarchical linear models, models of several time series, structural equation models, and other applications of the general linear model (e.g., the analysis of covariance). Further, many econometric models (e.g., investment, capital asset pricing, and production models) often have a common multiple equation structure that can be expressed as Equation 4.1 or Equation 4.2. See, for example, Green (1993, pp. 486–487) or Judge et al. (1988, pp. 448–449).

Most statistics textbooks discuss the validity of linear models or test sta-tistics in terms of the various assumptions concerning the error terms (ε_p) (Cook & Weisberg, 1999; Neter, Kutner, Nachtsheim, & Wasserman, 1996). For example, the usual OLS regression procedure assumes that the ε_p terms are independent and normally distributed with conditional expectation of zero and constant variance. Thus, in order to examine the properties of a system of statistical equations using Monte Carlo techniques, it is necessary to have an appropriate data generation procedure that allows for the specification of the distributional shapes and correlation structures of the error terms (such as ε in Equation 4.2). It is also desirable that this procedure be both computa-tionally efficient and general enough to allow for the simulation of a variety of settings, such as autocorrelation, heterogeneity of regression coefficients, heteroskedasticity, multicollinearity, nonnormality, and other violations of assumptions.

4.3 Methodology

Consider the linear system of T equations in Equation 4.2 more explicitly as

$$Y_p = \beta_{p0} + \sum_{i=1}^{k} \beta_{pi} X_{pi} + \sigma_p \varepsilon_p \tag{4.3}$$

where $p = 1, ..., T$ and $p \neq q$. We note that it is not necessary for each of the T equations to have the same number of independent variables (Xs).

Two nonnormal independent variables X_{pi} and X_{qi} in the system of Equation 4.3 are generated and correlated as described in Section 2.6 and demonstrated in Chapter 3 for fifth-order polynomials as

$$X_{pi} = \sum_{m=1}^{6} c_{mpi} W_{pi}^{m-1} \tag{4.4}$$

$$X_{qj} = \sum_{m=1}^{6} c_{mqj} W_{qj}^{m-1} \tag{4.5}$$

where W can follow either a standard normal, logistic, or uniform distribu-tion. Two nonnormal error terms (e.g., ε_p and ε_q) are generated and correlated in the same manner as in Equation 4.4 and Equation 4.5.

The dependent variables Y_p for the system are generated from the right-hand sides of the T equations in Equation 4.3. The method used to correlate Y_p with the k independent variables in the p-th equation is based on the following property:

Property 4.1

If the independent variables X_{pi} and X_{pj} have zero means, unit variances, correlation of $\rho_{X_{pi}X_{pj}}$, based on Equation 4.4 and Equation 4.5, and $\text{cov}(\varepsilon_p, X_{pi}) = 0$, then the correlation between Y_p and X_{pi} in Equation 4.3 is

$$\rho_{Y_pX_{pi}} = \frac{\beta_{pi} + \sum_{pj \neq pi} \beta_{pj}\rho_{X_{pi}X_{pj}}}{\sqrt{\sigma_p^2 + \sum_{pi} \beta_{pi}^2 + 2\sum_{pj \neq pi} \beta_{pi}\beta_{pj}\rho_{X_{pi}X_{pj}}}} \tag{4.6}$$

Equation 4.6 can be shown by using the definition of the product-moment of correlation and straightforward manipulations using the algebra of expectations.

The coefficients β_{pi} in Equation 4.3 are determined by simultaneously solving a system of k_p equations of the form of Equation 4.6. Specifically, the equations in this system specify pairwise correlations of $\rho_{Y_pX_{pi}}$ on the left-hand sides. The prespecified values of $\rho_{X_{pi}X_{pj}}$ and σ_p are substituted into the right-hand sides. The solutions for the coefficients β_{pi} are determined by numerically solving this system.

Remark 4.1

If the numerical solutions of all β_{pi} are finite real numbers, then the $(1+k_p) \times (1+k_p)$ correlation matrix associated with Y_p and all X_{pi} is sufficiently positive definite.

Given solved values of the coefficients β_{pi}, correlations between the independent variables, correlations between the error terms, and the scalar terms for the system in Equation 4.3, other correlations that may be of interest, such as $\rho_{Y_pY_q}$, $\rho_{Y_qX_{pi}}$, and $\rho_{Y_p(\sigma_q\varepsilon_q)}$, can be determined by evaluating

$$\rho_{Y_qX_{pi}} = \frac{\sum_{qi} \beta_{qi}\rho_{X_{pi}X_{qi}}}{\sqrt{\sigma_q^2 + \sum_{qi} \beta_{qi}^2 + 2\sum_{qj \neq qi} \beta_{qi}\beta_{qj}\rho_{X_{qi}X_{qj}}}} \tag{4.7}$$

$$\rho_{Y_pY_q} = \frac{\sigma_p\sigma_q\rho_{\varepsilon_p\varepsilon_q} + \sum_{pi}\sum_{qi} \beta_{pi}\beta_{qi}\rho_{X_{pi}X_{qi}}}{\sqrt{\sigma_p^2 + \sum_{pi} \beta_{pi}^2 + 2\sum_{pj \neq pi} \beta_{pi}\beta_{pj}\rho_{X_{pi}X_{pj}}}\sqrt{\sigma_q^2 + \sum_{qi} \beta_{qi}^2 + 2\sum_{qj \neq qi} \beta_{qi}\beta_{qj}\rho_{X_{qi}X_{qj}}}} \tag{4.8}$$

$$\rho_{Y_p(\sigma_q\varepsilon_q)} = \frac{\sigma_p\rho_{\varepsilon_p\varepsilon_q}}{\sqrt{\sigma_p^2 + \sum_{pi} \beta_{pi}^2 + 2\sum_{pj \neq pi} \beta_{pi}\beta_{pj}\rho_{X_{pi}X_{pj}}}} \tag{4.9}$$

The derivations of Equations 4.7, 4.8, and 4.9 are analogous to Equation 4.6. Note that if we were to set $\rho_{\varepsilon_p \varepsilon_q} = 1$ in 4.9, then this would yield the correlation between Y_p and its error term $\sigma_p \varepsilon_p$, that is, $\rho_{Y_p (\sigma_p \varepsilon_p)}$.

4.4 Numerical Example and Monte Carlo Simulation

Suppose we desire the system

$$y = x\beta + \sigma\varepsilon \qquad (4.10)$$

where $p = 1, ..., T = 4; k = 4$ for all $p = 1, ..., 4$. If $N = 10$, for example, then y and ε would have dimension (40×1), x is (40×16), and β is (16×1). The specified parameters for this numerical example are summarized in Tables 4.1–4.7 for fifth-order normal-, logistic-, and uniform-based power method polynomials. Specifically, Tables 4.1–4.3 give the specified correlations between the independent variables (x), the error terms (ε), and the correlations between the dependent variables (y) and x, respectively. Tables 4.4 and 4.5 give the specified cumulants and their associated polynomial coefficients for x and ε. Table 4.6 gives the solutions for the beta coefficients (β) using Equation 4.6

TABLE 4.1

Specified Correlations for the Independent Variables X_{pi}

X_{pi}	11	12	13	14	21	22	23	24	31	32	33	34	41	42	43	44
11	1	.2	.2	.2	.3	.3	.3	.3	.3	.3	.3	.3	.3	.3	.3	.3
12	.2	1	.2	.2	.3	.3	.3	.3	.3	.3	.3	.3	.3	.3	.3	.3
13	.2	.2	1	.2	.3	.3	.3	.3	.3	.3	.3	.3	.3	.3	.3	.3
14	.2	.2	.2	1	.3	.3	.3	.3	.3	.3	.3	.3	.3	.3	.3	.3
21	.3	.3	.3	.3	1	.4	.4	.4	.5	.5	.5	.5	.5	.5	.5	.5
22	.3	.3	.3	.3	.4	1	.4	.4	.5	.5	.5	.5	.5	.5	.5	.5
23	.3	.3	.3	.3	.4	.4	1	.4	.5	.5	.5	.5	.5	.5	.5	.5
24	.3	.3	.3	.3	.4	.4	.4	1	.5	.5	.5	.5	.5	.5	.5	.5
31	.3	.3	.3	.3	.5	.5	.5	.5	1	.6	.6	.6	.7	.7	.7	.7
32	.3	.3	.3	.3	.5	.5	.5	.5	.6	1	.6	.6	.7	.7	.7	.7
33	.3	.3	.3	.3	.5	.5	.5	.5	.6	.6	1	.6	.7	.7	.7	.7
34	.3	.3	.3	.3	.5	.5	.5	.5	.6	.6	.6	1	.7	.7	.7	.7
41	.3	.3	.3	.3	.5	.5	.5	.5	.7	.7	.7	.7	1	.8	.8	.8
42	.3	.3	.3	.3	.5	.5	.5	.5	.7	.7	.7	.7	.8	1	.8	.8
43	.3	.3	.3	.3	.5	.5	.5	.5	.7	.7	.7	.7	.8	.8	1	.8
44	.3	.3	.3	.3	.5	.5	.5	.5	.7	.7	.7	.7	.8	.8	.8	1

Note: The column and row headings are subscripts identifying the position of the *pi*-th independent variable.

TABLE 4.2

Specified Correlations for the Error Terms ε_p

	ε_1	ε_2	ε_3	ε_4
ε_1	1	0.60	0.50	0.40
ε_2	0.60	1	0.50	0.40
ε_3	0.50	0.50	1	0.40
ε_4	0.40	0.40	0.40	1

TABLE 4.3

Specified Correlations for the Dependent Variable Y_p and Independent Variables X_{pi}

Equation (p)	X_{p1}	X_{p2}	X_{p3}	X_{p4}
$Y_{p=1}$	0.20	0.20	0.10	0.10
$Y_{p=2}$	0.40	0.40	0.20	0.20
$Y_{p=3}$	0.60	0.60	0.40	0.40
$Y_{p=4}$	0.80	0.80	0.60	0.60

TABLE 4.4

Standardized Cumulants Associated with ε_p and X_{pi} for the Normal-, Logistic-, and Uniform-Based Polynomials

Normal	γ_3	γ_4	γ_5	γ_6
ε_1, X_{1i}	0.5	0.375	0.46875	1
ε_2, X_{2i}	1	1.5	3	7.5
ε_3, X_{3i}	$\sqrt{3}$	4.5	15.588	67.5
ε_4, X_{4i}	$\sqrt{5}$	7.5	33.541	187.5
Logistic				
ε_1, X_{1i}	1	3	9	300
ε_2, X_{2i}	2	12	72	4800
ε_3, X_{3i}	3.5	36.5	385.875	45018.75
ε_4, X_{4i}	5	75	1125	187500
Uniform				
ε_1, X_{1i}	0.25	0.125	−0.875	−3.5
ε_2, X_{2i}	0.5	0.25	−1.75	−7
ε_3, X_{3i}	0.75	0.375	−2.625	−10
ε_4, X_{4i}	1	0.5	−3.5	−14

TABLE 4.5

Polynomial Coefficients for Independent Variables X_p and the Error Terms ε_p Associated with the p-th Equation

Normal	c_{1p}	c_{2p}	c_{3p}	c_{4p}	c_{5p}	c_{6p}
$p = 1$	−0.081896	0.992059	0.080737	−0.000545	0.000386	0.0001763
$p = 2$	−0.163968	0.950784	0.165391	0.007635	−0.000474	0.0000144
$p = 3$	−0.272732	0.850315	0.280242	0.024278	−0.002503	0.0001115
$p = 4$	−0.336017	0.752533	0.350033	0.042730	−0.004672	0.0001728
Logistic						
$p = 1$	−0.113448	0.974310	0.119946	0.001045	−0.001618	0.0001010
$p = 2$	−0.175167	0.848122	0.184092	0.023275	−0.002125	0.0000959
$p = 3$	−0.239383	0.690902	0.249991	0.045778	−0.002526	0.0001185
$p = 4$	−0.288624	0.548385	0.300275	0.062593	−0.002774	0.0001549
Uniform						
$p = 1$	−0.098739	0.826890	0.140735	−0.291327	−0.023331	0.1722165
$p = 2$	−0.153567	0.311246	0.177731	0.334038	−0.013424	0.0095045
$p = 3$	−0.253093	0.395892	0.317051	0.217492	−0.035532	0.0394731
$p = 4$	−0.359484	0.477757	0.454261	0.118792	−0.052653	0.0601659

and the correlations listed in Tables 4.1 and 4.3, where the scale parameters $\boldsymbol{\sigma}$ are set to $\boldsymbol{\sigma} = (\sigma_1 = 0.50, \sigma_2 = 1.0, \sigma_3 = 2.0, \sigma_4 = 3.0)$ to induce heteroskedasticity. Last, Table 4.7 gives the intermediate correlations for the independent variables and error terms for the specified correlations in Tables 4.1 and 4.2.

The data generation procedure begins by following the steps as outlined in Section 2.6. Specifically, correlation matrices for the independent variables and error terms with the dimensions in Tables 4.1 and 4.2 are created with the intermediate correlations listed in Table 4.7 as their entries. The intermediate correlations were obtained using Equation 2.59 for the normal-based system and Equation 2.55 for the logistic- and uniform-based systems. The matrices are then separately factored and the results from these factorizations

TABLE 4.6

The β Coefficients for Generating the Specified Correlations between the Dependent Variable Y_p and the Independent Variables X_{pi} in Table 4.3

Equation (p)	β_{p1}	β_{p2}	β_{p3}	β_{p4}
$p = 1$	0.0809574	0.0809574	0.0161915	0.0161915
$p = 2$	0.3454030	0.3454030	−0.0345403	−0.0345403
$p = 3$	1.1633501	1.1633501	−0.1938917	−0.1938917
$p = 4$	4.4790534	4.4790534	−1.8662722	−1.8662722

TABLE 4.7

Specified Correlations and Intermediate (Inter-) Correlations for the Independent Variables X_{pi} and the Error Terms ε_p in Tables 4.1 and 4.2

Variables	Specified	Internormal	Interlogistic	Interuniform
X_{1i}	0.20	0.202229	0.207662	0.202376
X_{1i}, X_{2i}	0.30	0.307813	0.321127	0.307225
X_{1i}, X_{3i}	0.30	0.321681	0.343031	0.310115
X_{1i}, X_{4i}	0.30	0.335138	0.370997	0.315806
X_{2i}	0.40	0.412927	0.434899	0.413690
X_{2i}, X_{3i}	0.50	0.528655	0.560245	0.519543
X_{2i}, X_{4i}	0.50	0.545154	0.591414	0.527286
X_{3i}	0.60	0.634177	0.670895	0.623190
X_{3i}, X_{4i}	0.70	0.739193	0.778702	0.728910
X_{4i}	0.80	0.832181	0.862722	0.826260
$\varepsilon_1, \varepsilon_2$	0.60	0.610498	0.627460	0.612419
$\varepsilon_1, \varepsilon_3$	0.50	0.530871	0.558045	0.515412
$\varepsilon_1, \varepsilon_4$	0.40	0.443952	0.486344	0.420309
$\varepsilon_2, \varepsilon_3$	0.50	0.528655	0.560245	0.519543
$\varepsilon_2, \varepsilon_4$	0.40	0.441448	0.486354	0.423458
$\varepsilon_3, \varepsilon_4$	0.40	0.447101	0.497608	0.425020

are used to produce standard normal deviates (e.g., Equation 2.56). For the logistic- and uniform-based polynomials, standard normal deviates are converted using Equations 2.53 and 2.54. The standard normal, logistic, and uniform deviates are subsequently transformed using fifth-order polynomials with the power constants listed in Table 4.5 to produce the independent variables and error terms with their specified correlations and shapes. The three systems (normal, logistic, uniform) are then generated using equations of the form in Equation 4.3 with the β coefficients listed in Table 4.6 and the specified scalar terms σ.

To evaluate the procedure, distributions with the specified cumulants and correlations associated with x and ε were generated. Single samples of size 3 million were drawn for each distribution using the normal-, logistic-, and uniform-based polynomial transformations. A representative subset of all specified correlations and standardized cumulants was estimated based on these samples. Further, estimates were also computed for the parameters associated with Equations 4.7–4.9. The estimates are reported in Tables 4.8–4.16 for the correlations and cumulants associated with the normal-, logistic-, and uniform-based systems. Inspection of these tables indicates that the procedure generates estimates of correlations and standardized cumulants that are very close to the parameters.

TABLE 4.8

Specified and Estimated (Est.) Correlations between the Independent Variables X_{pi} and X_{pj}

Variables	Specified Correlation	Est. Normal	Est. Logistic	Est. Uniform
X_{11}, X_{12}	0.20	0.200	0.199	0.200
X_{11}, X_{13}	0.20	0.201	0.199	0.199
X_{11}, X_{14}	0.20	0.200	0.199	0.200
X_{12}, X_{13}	0.20	0.201	0.199	0.200
X_{12}, X_{14}	0.20	0.201	0.201	0.200
X_{13}, X_{14}	0.20	0.201	0.200	0.201
X_{21}, X_{22}	0.40	0.401	0.401	0.400
X_{21}, X_{23}	0.40	0.401	0.399	0.399
X_{21}, X_{24}	0.40	0.400	0.400	0.400
X_{22}, X_{23}	0.40	0.400	0.401	0.399
X_{22}, X_{24}	0.40	0.400	0.400	0.400
X_{23}, X_{24}	0.40	0.400	0.399	0.400
X_{31}, X_{32}	0.60	0.600	0.600	0.600
X_{31}, X_{33}	0.60	0.601	0.600	0.600
X_{31}, X_{34}	0.60	0.601	0.600	0.600
X_{32}, X_{33}	0.60	0.600	0.599	0.599
X_{32}, X_{34}	0.60	0.600	0.599	0.600
X_{33}, X_{34}	0.60	0.601	0.599	0.600
X_{41}, X_{42}	0.80	0.801	0.800	0.800
X_{41}, X_{43}	0.80	0.800	0.801	0.800
X_{41}, X_{44}	0.80	0.800	0.800	0.800
X_{42}, X_{43}	0.80	0.800	0.800	0.800
X_{42}, X_{44}	0.80	0.800	0.800	0.800
X_{43}, X_{44}	0.80	0.800	0.801	0.800

4.5 Some Additional Comments

The procedure described above may also be used for other systems of statistical models based on the general linear model (GLM). For example, the method could be used to generate T independent models to investigate the properties (e.g., Type I error and power) of competing nonparametric tests in the context of the analysis of covariance (ANCOVA) or repeated measures. In terms of ANCOVA, one attractive feature of the proposed method is that it has an advantage over other competing algorithms to the extent that it allows for

TABLE 4.9

Specified and Estimated (Est.) Correlations between X_{pi} and X_{qi}

Variables	Specified Correlation	Est. Normal	Est. Logistic	Est. Uniform
X_{11}, X_{21}	0.30	0.300	0.299	0.299
X_{11}, X_{24}	0.30	0.300	0.299	0.299
X_{12}, X_{22}	0.30	0.300	0.300	0.299
X_{12}, X_{23}	0.30	0.300	0.300	0.300
X_{13}, X_{23}	0.30	0.300	0.301	0.299
X_{13}, X_{22}	0.30	0.300	0.301	0.300
X_{14}, X_{24}	0.30	0.302	0.301	0.300
X_{14}, X_{21}	0.30	0.299	0.300	0.300
X_{31}, X_{21}	0.50	0.501	0.501	0.500
X_{31}, X_{24}	0.50	0.500	0.500	0.500
X_{32}, X_{22}	0.50	0.500	0.501	0.500
X_{32}, X_{23}	0.50	0.500	0.500	0.500
X_{33}, X_{23}	0.50	0.500	0.500	0.499
X_{33}, X_{22}	0.50	0.500	0.501	0.500
X_{34}, X_{24}	0.50	0.500	0.501	0.500
X_{34}, X_{21}	0.50	0.500	0.499	0.500
X_{31}, X_{41}	0.70	0.701	0.700	0.700
X_{31}, X_{44}	0.70	0.701	0.700	0.700
X_{32}, X_{42}	0.70	0.700	0.700	0.699
X_{32}, X_{43}	0.70	0.699	0.701	0.699
X_{33}, X_{43}	0.70	0.701	0.700	0.700
X_{33}, X_{42}	0.70	0.701	0.699	0.700
X_{34}, X_{44}	0.70	0.700	0.700	0.700
X_{34}, X_{41}	0.70	0.700	0.700	0.700

the creation of distributions with unequal regression slopes while maintaining equal variances. This can be demonstrated by inspecting Equation 4.3, where the slope coefficient(s) could change (i.e., be made unequal) while the error terms remain unchanged. Subsequent to any changes made to the slope coefficients, the variate (or dependent variable) and covariate correlations can be determined using Equation 4.6.

Many other applications of the proposed procedure to the GLM are possible. From the GLM perspective, the dependent variables Y_p could represent the same variable collected under T different conditions or at time points $1, \ldots, T$. In either case, the independent variables X_{pi} could represent static covariates (e.g., preexisting ability measures often used in ANCOVA models). On the other hand, the independent variables X_{pi} could also be different for each of the T equations and may be used to represent time-varying covariates (i.e., some variables measured over the T periods along with Y_p).

TABLE 4.10

Specified and Estimated (Est.) Correlations between Y_p and X_{pi}

Variables	Specified Correlation	Est. Normal	Est. Logistic	Est. Uniform
Y_1, X_{11}	0.20	0.200	0.200	0.200
Y_1, X_{12}	0.20	0.200	0.200	0.199
Y_1, X_{13}	0.10	0.099	0.100	0.100
Y_1, X_{14}	0.10	0.099	0.101	0.101
Y_2, X_{21}	0.40	0.399	0.400	0.399
Y_2, X_{22}	0.40	0.399	0.399	0.399
Y_2, X_{23}	0.20	0.200	0.200	0.199
Y_2, X_{24}	0.20	0.200	0.200	0.201
Y_3, X_{31}	0.60	0.600	0.600	0.600
Y_3, X_{32}	0.60	0.600	0.599	0.600
Y_3, X_{33}	0.40	0.400	0.399	0.399
Y_3, X_{34}	0.40	0.400	0.400	0.400
Y_4, X_{41}	0.80	0.801	0.800	0.800
Y_4, X_{42}	0.80	0.801	0.799	0.800
Y_4, X_{43}	0.60	0.601	0.600	0.600
Y_4, X_{44}	0.60	0.600	0.600	0.600

In the context of repeated measures, the method could also be used to allow the simulation of repeated measures data of nonspherical structures with nonnormal errors and covariates. It should also be noted that with nonnormal error terms one could specify all of these distributions to be the same. This assumption is implicit in parametric analyses of repeated measures data because normal error populations are assumed.

The procedure could also be applied in terms of time series analysis. Specifically, the procedure could be used to model instrumental variables that address one of the problems that certain autoregressive models

TABLE 4.11

Specified and Estimated (Est.) Correlations between the Error Terms ε_p and ε_q

Variables	Specified Correlation	Est. Normal	Est. Logistic	Est. Uniform
$\varepsilon_1, \varepsilon_2$	0.60	0.600	0.599	0.600
$\varepsilon_1, \varepsilon_3$	0.50	0.500	0.499	0.500
$\varepsilon_1, \varepsilon_4$	0.40	0.400	0.399	0.399
$\varepsilon_2, \varepsilon_3$	0.50	0.500	0.499	0.500
$\varepsilon_2, \varepsilon_4$	0.40	0.400	0.398	0.399
$\varepsilon_3, \varepsilon_4$	0.40	0.400	0.398	0.400

TABLE 4.12

Population and Estimated (Est.) Correlations between the Dependent Variables (Y) and the Error Terms (ε)

Variables	Population Correlation	Est. Normal	Est. Logistic	Est. Uniform
Y_1, ε_1	0.965	0.965	0.965	0.965
Y_1, ε_2	0.579	0.579	0.578	0.579
Y_1, ε_3	0.483	0.482	0.481	0.482
Y_1, ε_4	0.386	0.386	0.384	0.385
Y_2, ε_1	0.526	0.526	0.526	0.527
Y_2, ε_2	0.877	0.877	0.877	0.877
Y_2, ε_3	0.439	0.438	0.437	0.438
Y_2, ε_4	0.351	0.351	0.349	0.350
Y_3, ε_1	0.368	0.368	0.369	0.368
Y_3, ε_2	0.368	0.368	0.368	0.368
Y_3, ε_3	0.737	0.737	0.738	0.737
Y_3, ε_4	0.295	0.295	0.293	0.295
Y_4, ε_1	0.189	0.188	0.189	0.188
Y_4, ε_2	0.189	0.188	0.189	0.188
Y_4, ε_3	0.189	0.188	0.189	0.189
Y_4, ε_4	0.473	0.471	0.474	0.473
Y_1, Y_2	0.569	0.569	0.569	0.569
Y_1, Y_3	0.436	0.435	0.436	0.435
Y_1, Y_4	0.275	0.275	0.275	0.274
Y_2, Y_3	0.518	0.518	0.517	0.518
Y_2, Y_4	0.391	0.390	0.390	0.389
Y_3, Y_4	0.551	0.551	0.550	0.551

(e.g., the adaptive expectations model) have where the dependent variable from a preceding time period (Y_p) is included as an independent variable in the subsequent (q-th) period. As such, the vectors Y_p and ε_q are usually correlated. Using the suggested procedures above, a Monte Carlo study could be arranged to simulate nonnormal "proxies" correlated at various levels between Y_p and ε_q.

The application of the proposed method to the GLM is flexible and has the potential to simulate other types of models where the error distributions may change over time. For example, data sets with repeated measures often have mistimed measures or missing data. Thus, the proposed method

TABLE 4.13

Population and Estimated (Est.) Correlations between the Dependent Variables (Y_p) and the Independent Variables (X_{qi})

Variables	Population Correlation	Est. Normal	Est. Logistic	Est. Uniform
Y_1, X_{21}	0.1125	0.112	0.112	0.112
Y_1, X_{24}	0.1125	0.113	0.114	0.113
Y_1, X_{31}	0.1125	0.112	0.113	0.112
Y_1, X_{34}	0.1125	0.112	0.113	0.111
Y_1, X_{41}	0.1125	0.112	0.113	0.112
Y_1, X_{44}	0.1125	0.112	0.113	0.112
Y_2, X_{11}	0.164	0.164	0.163	0.163
Y_2, X_{14}	0.164	0.164	0.164	0.164
Y_2, X_{31}	0.273	0.526	0.526	0.527
Y_2, X_{34}	0.273	0.877	0.877	0.877
Y_2, X_{41}	0.273	0.438	0.437	0.438
Y_2, X_{44}	0.273	0.351	0.349	0.350
Y_3, X_{11}	0.214	0.214	0.213	0.214
Y_3, X_{14}	0.214	0.214	0.214	0.215
Y_3, X_{21}	0.357	0.357	0.357	0.356
Y_3, X_{24}	0.357	0.357	0.357	0.357
Y_3, X_{41}	0.500	0.500	0.500	0.500
Y_3, X_{44}	0.500	0.499	0.500	0.499
Y_4, X_{11}	0.247	0.247	0.246	0.246
Y_4, X_{14}	0.247	0.248	0.247	0.248
Y_4, X_{21}	0.412	0.412	0.411	0.411
Y_4, X_{24}	0.412	0.412	0.411	0.412
Y_4, X_{31}	0.576	0.578	0.576	0.577
Y_4, X_{34}	0.576	0.577	0.576	0.576

could be used to compare and contrast the GLM with generalized estimating equations (GEEs) or hierarchical linear models (HLMs)—procedures that are often considered preferable to standard univariate or multivariate OLS procedures.

It may also be reasonable to consider a model where the correlation structure is a function of time between the observations. Thus, data could also be generated using the proposed procedure for Monte Carlo studies involving dynamic regression models that have distributed lags or moving averages.

TABLE 4.14

Estimates of the First Four Standardized Cumulants Associated with the Normal-Based Polynomials Listed in Table 4.4

	$\hat{\gamma}_1$	$\hat{\gamma}_2$	$\hat{\gamma}_3$	$\hat{\gamma}_4$
ε_1	−0.0002	0.2503	0.4994	0.3694
ε_2	−0.0007	1.0008	1.0023	1.5111
ε_3	0.0005	4.0022	1.7333	4.5125
ε_4	0.0002	8.9802	2.2252	7.4160
X_{11}	−0.0008	1.0013	0.4999	0.3734
X_{12}	0.0005	1.0024	0.5051	0.3908
X_{13}	0.0005	0.9991	0.5010	0.3813
X_{14}	−0.0003	1.0017	0.5001	0.3779
X_{21}	0.0006	1.0003	0.9996	1.4958
X_{22}	−0.0005	0.9980	0.9995	1.4976
X_{23}	0.0003	1.0000	1.0009	1.5057
X_{24}	0.0001	1.0003	1.0002	1.4998
X_{31}	0.0002	1.0025	1.7393	4.5387
X_{32}	0.0003	1.0000	1.7372	4.5491
X_{33}	0.0008	1.0029	1.7388	4.5423
X_{34}	0.0012	1.0038	1.7366	4.5222
X_{41}	0.0009	1.0042	2.2496	7.6283
X_{42}	0.0012	1.0036	2.2423	7.5529
X_{43}	0.0001	0.9999	2.2430	7.6156
X_{44}	0.0008	1.0028	2.2445	7.5802

Note: The error terms $(\sigma_i \varepsilon_i)$ were scaled such that their variances are $\sigma_1^2 = 0.25$, $\sigma_2^2 = 1$, $\sigma_3^2 = 4$, and $\sigma_4^2 = 9$.

4.6 Simulating Intraclass Correlation Coefficients

In this section the focus is turned to simulating nonnormal distributions with specified intraclass correlation coefficients (ICCs). ICCs are commonly used statistics in a variety of settings. Some examples include assessing reliability (Bartko, 1976; Shrout & Fleiss, 1979), generalizability studies (Cronbach, Gleser, Nanda, & Rajaratnam, 1972), longitudinal studies involving variation among twins in heritability studies (Christian, Yu, Slemenda, & Johnston, 1989; Hanisch, Dittmar, Hohler, & Alt, 2004), reproducibility studies (Giraudeau & Mary, 2001), smoking prevention studies (Murray et al., 1994; Siddiqui, Hedeker, Flay, & Hu, 1996), and survey research with

TABLE 4.15

Estimates of the First Four Standardized Cumulants Associated with the Logistic-Based Polynomials Listed in Table 4.4

	$\hat{\gamma}_1$	$\hat{\gamma}_2$	$\hat{\gamma}_3$	$\hat{\gamma}_4$
ε_1	−0.0001	0.2502	1.0008	2.9586
ε_2	−0.0001	1.0015	1.9880	12.4025
ε_3	−0.0015	4.0110	3.5379	38.7784
ε_4	−0.0008	9.0560	5.2822	78.0400
X_{11}	−0.0005	0.9997	0.9890	3.0065
X_{12}	−0.0001	1.0010	1.0074	3.0282
X_{13}	0.0005	1.0013	1.0030	2.9496
X_{14}	−0.0005	0.9981	1.0023	2.9487
X_{21}	0.0001	1.0002	1.9688	12.4847
X_{22}	0.0000	0.9972	2.0138	11.6576
X_{23}	−0.0004	1.0016	1.8835	13.9371
X_{24}	0.0011	1.0021	1.9832	13.3762
X_{31}	0.0005	1.0013	3.5591	35.2594
X_{32}	−0.0001	0.9965	3.5561	38.0802
X_{33}	0.0004	1.0004	3.5094	33.2557
X_{34}	0.0003	1.0010	3.5837	34.3692
X_{41}	0.0005	1.0051	5.1968	81.4517
X_{42}	0.0003	0.9975	4.9300	66.9822
X_{43}	0.0002	1.0004	5.1675	76.0770
X_{44}	0.0006	0.9992	4.8407	67.9053

Note: The error terms $(\sigma_i \varepsilon_i)$ were scaled such that their variances are $\sigma_1^2 = 0.25$, $\sigma_2^2 = 1$, $\sigma_3^2 = 4$, and $\sigma_4^2 = 9$.

clustered data (Rowe, Lama, Onikpo, & Deming, 2002). Although there are several different kinds of ICCs, what these indices have in common is that they provide a measure of homogeneity among the analytical units under study.

The properties of ICC statistics most often used are derived under classical normal curve theory and are consistent (not unbiased) estimators of their associated population parameters. As such, because data sets are frequently nonnormal (e.g., Micceri, 1989; Pearson & Please, 1975), it may be desirable to investigate the properties of ICCs (e.g., McGraw & Wong, 1996; Wong & McGraw, 2005) under a variety of conditions. The usual approach used to carry out such investigations is Monte Carlo techniques.

TABLE 4.16

Estimates of the First Four Standardized Cumulants Associated with the Uniform-Based Polynomials Listed in Table 4.4

	$\hat{\gamma}_1$	$\hat{\gamma}_2$	$\hat{\gamma}_3$	$\hat{\gamma}_4$
ε_1	−0.0001	0.2498	0.2480	0.1232
ε_2	0.0003	1.0014	0.4983	0.2437
ε_3	0.0008	3.9966	0.7502	0.3755
ε_4	0.0000	8.9923	1.0012	0.5038
X_{11}	−0.0004	1.0005	0.2508	0.1242
X_{12}	−0.0003	0.9991	0.2500	0.1284
X_{13}	0.0009	1.0009	0.2497	0.1250
X_{14}	0.0006	1.0007	0.2495	0.1250
X_{21}	−0.0003	1.0001	0.5000	0.2479
X_{22}	−0.0002	0.9983	0.5000	0.2509
X_{23}	0.0004	0.9990	0.4993	0.2520
X_{24}	−0.0008	1.0001	0.4991	0.2456
X_{31}	−0.0009	0.9983	0.7510	0.3783
X_{32}	−0.0003	0.9983	0.7501	0.3778
X_{33}	−0.0003	0.9992	0.7513	0.3801
X_{34}	−0.0009	0.9989	0.7513	0.3793
X_{41}	−0.0009	0.9993	1.0014	0.5049
X_{42}	−0.0011	0.9986	1.0012	0.5055
X_{43}	−0.0012	0.9985	1.0009	0.5060
X_{44}	−0.0009	0.9990	1.0013	0.5044

Note: The error terms $(\sigma_i \varepsilon_i)$ were scaled such that their variances are $\sigma_1^2 = 0.25$, $\sigma_2^2 = 1$, $\sigma_3^2 = 4$, and $\sigma_4^2 = 9$.

4.7 Methodology

The linear model we will consider for simulating nonnormal distributions with specified ICCs is

$$Y_{ij} = \mu + \pi_i + \tau_j + (\pi\tau)_{ij} + \varepsilon_{ij} \tag{4.11}$$

Without loss of generality, the populations associated with Y_{ij} are assumed to have an overall mean of $\mu = 0$ and unit variances. The term π_i is a parameter that denotes the effect on the i-th object, τ_j is a parameter that denotes the effect on the j-th class (i.e., $\tau_j = \mu_j$ since $\mu = 0$), $(\pi\tau)_{ij}$ is a parameter that denotes the interaction between the i-th object and the j-th class, and ε_{ij} is the stochastic error component, where $i = 1, \ldots, n$; $j = 1, \ldots, k$, and the total number of observations is $nk = N$.

					Class			
		1	2	3	...	*j*	...	*k*
Object								
1								
2								
3								
⋮								
i								
⋮								
N								

FIGURE 4.1
Schematic of the design associated with the linear model in Equation 4.11.

A schematic and an ANOVA summary associated with Equation 4.11 are given in Figure 4.1 and Table 4.17. The computational formulae for the ICC sample statistics considered herein are given in Table 4.18. The four ICCs in Table 4.18 are perhaps the most commonly reported in the applied literature and were first collectively reviewed in the often-cited paper by Shrout and Fleiss (1979). For example, ICC_4 is the index that is also referred to as Cronbach's alpha (1951).

The expected values of the mean squares in Table 4.18 for k standardized populations of Y_{ij} are

$$E[MS_B] = E[\overline{var} + (k-1)\overline{cov}] = 1 + (k-1)\bar{\rho} \tag{4.12}$$

$$E[MS_C] = (1 - \bar{\rho}) + n\sigma_\tau^2 \tag{4.13}$$

$$E[MS_{C\times O}] = E[\overline{var} - \overline{cov}] = 1 - \bar{\rho} \tag{4.14}$$

The terms $\overline{var}$ and $\overline{cov}$ in Equations 4.12 and 4.14 are the means of the variances and covariances from the $k \times k$ variance-covariance matrix. The

TABLE 4.17

ANOVA Summary for the Linear Model in Equation 4.11 and Figure 4.1

Source	df	Sums of Squares	Mean Squares
Between objects	$n-1$	SS_B	$MS_B = \dfrac{SS_B}{n-1}$
Within objects	$n(k-1)$	SS_W	$MS_W = \dfrac{SS_W}{n(k-1)}$
Class	$k-1$	SS_C	$MS_C = \dfrac{SS_C}{k-1}$
Class × objects	$(k-1)(n-1)$	$SS_{C\times O}$	$MS_{C\times O} = \dfrac{SS_{C\times O}}{(k-1)(n-1)}$
Total	$N-1$		

TABLE 4.18

Sample Statistics for the Intraclass Correlations ($ICC_{t=1,2,3,4}$) Considered

Classes	Random		Fixed	
Single class	$ICC_1 = \dfrac{MS_B - MS_{C\times O}}{MS_B + (k-1)MS_{C\times O} + k(MS_C - MS_{C\times O})/n}$		$ICC_2 = \dfrac{MS_B - MS_{C\times O}}{MS_B + (k-1)MS_{C\times O}}$	
All classes	$ICC_3 = \dfrac{MS_B - MS_{C\times O}}{MS_B + (MS_C - MS_{C\times O})/n}$		$ICC_4 = \dfrac{MS_B - MS_{C\times O}}{MS_B}$	

parameter $\bar{\rho} = \Sigma \rho_{ij}/k$ in Equations 4.12–4.14 is the population mean of the $k(k-1)/2$ correlations. The term σ_τ^2 in Equation 4.13 is the variance component associated with the effects on the k classes and is determined as $\sigma_\tau^2 = \Sigma \mu_i^2/(k-1)$. Substituting Equations 4.12–4.14 into the estimates of the mean squares in Table 4.18 and simplifying gives the ICC population parameters that are given in Table 4.19. For the two cases in Table 4.19, where the classes are assumed to be random, σ_τ^2 can be expressed as a function of $\bar{\rho}$ and either ρ_{I_1} or ρ_{I_3} as

$$\sigma_\tau^2 = \frac{\bar{\rho} - \rho_{I_1}}{\rho_{I_1}} \tag{4.15}$$

$$\sigma_\tau^2 = \bar{\rho}(1 + k(1/\rho_{I_3} - 1)) - 1 \tag{4.16}$$

As in Equation 4.4, nonnormal distributions for Y_{ij} in Equation 4.11 will be generated using a fifth-order normal-based transformation. The j-th variate is summarized as

$$Y_{ij} = \sum_{m=1}^{6} c_{mj} Z_{ij}^{m-1} \tag{4.17}$$

Using the equations above, in this section, and the method described in Section 2.6 for simulating correlated data, an outline for conducting a Monte Carlo study is subsequently described below.

TABLE 4.19

Parameters for the ICCs (ρ_{I_t}) Considered in Table 4.18

Class	Random	Fixed
Single class	$\rho_{I_1} = \dfrac{\bar{\rho}}{1 + \sigma_\tau^2}$	$\rho_{I_2} = \bar{\rho}$
All classes	$\rho_{I_3} = \dfrac{k\bar{\rho}}{1 + (k-1)\bar{\rho} + \sigma_\tau^2}$	$\rho_{I_4} = \dfrac{k\bar{\rho}}{1 + (k-1)\bar{\rho}}$

1. Specify k classes, the standardized cumulants γ_{mj} for Y_{ij}, and compute the polynomial coefficients c_{mj} associated with Equation 4.17. See Section 2.4. We note that it is not necessary to have each of the k classes follow the same distribution.

2. Specify a $k \times k$ correlation matrix such that Y_{ii} and Y_{ij} have intercorrelations $\rho_{Y_{ii}Y_{ij}}$ that average to a specified population mean of $\rho_{I_2} = \bar{\rho} = \Sigma \rho_{Y_{ii}Y_{ij}}/k$.

3. Determine the intermediate correlation matrix $(\rho_{Z_{ii}Z_{ij}})$ using Equation 2.59 with the specified coefficients c_{mj} and correlations $\rho_{Y_{ii}Y_{ij}}$ from Steps 1 and 2.

4. Specify the other ICC (e.g., ρ_{I_1}) of interest and determine the variance component parameter σ_τ^2 by evaluating Equations 4.15 or 4.16 (if needed).

5. Select shift parameters μ_j such that σ_τ^2 from Step 4 is $\sigma_\tau^2 = \Sigma \mu_j^2/(k-1)$ and where $\Sigma \mu_j = \mu = 0$.

6. Determine the values of the other ICCs (e.g., ρ_{I_3} and ρ_{I_4}) by evaluating the formulae in Table 4.19 given k, $\bar{\rho}$, and σ_τ^2.

7. Factor the intermediate correlation matrix (e.g., a Cholesky factorization). Use the results from this factorization to generate standard normal random deviates correlated at the intermediate levels $\rho_{Z_{ii}Z_{ij}}$ as described in Section 2.6 and equations of the form in Equation 2.56.

8. Use equations of the form in Equation 4.17, the standard normal random deviates from Step 7, and the shift parameters μ_j determined in Step 5 to generate the k vectors of Y_{ij}.

4.8 Numerical Example and Monte Carlo Simulation

Consider the specified (non)normal distributions and correlations $\rho_{Y_{ii}Y_{ij}}$ given in Table 4.20. Table 4.21 gives a summary of the eight steps above with calculations for an example with $k = 3$ classes and where each class has the same (non)normal distribution. To confirm the calculations associated with this example, independent sample sizes of $n = 10, 30, 100$, and $1,000$ for each of the four distributions specified in Table 4.20 were generated for simulating the specified ICC parameters $\rho_{I_1} = 0.60$, $\rho_{I_2} = 0.70$, $\rho_{I_3} = 0.8181...$, and $\rho_{I_4} = 0.875$.

For each distribution and sample size, 25,000 ICC$_t$ statistics were generated using the formulae from Tables 4.17 and 4.18 for all $t = 1,...,4$. These statistics were subsequently transformed to (Hotelling, 1953)

$$z^* = z - \frac{3z + \text{ICC}_t}{4n} - \frac{23z + 33(\text{ICC}_t) - 5(\text{ICC}_t)^3}{96n^2} \tag{4.18}$$

where $z = 0.5 \ln(|1 + \text{ICC}_t|/|1 - \text{ICC}_t|)$ is the usual Fisher (1921) z transformation.

TABLE 4.20

Distributions and Other Specifications for the Simulation

Cumulants	Coefficients	Correlations	Probability Density Function
$\gamma_1 = 0$	$c_1 = 0$	$\rho_{Y_{i1}Y_{i2}} = 0.650$	
$\gamma_2 = 1$	$c_2 = 1$	$\rho_{Y_{i1}Y_{i3}} = 0.700$	
$\gamma_3 = 0$	$c_3 = 0$	$\rho_{Y_{i2}Y_{i3}} = 0.750$	
$\gamma_4 = 0$	$c_4 = 0$	$\rho_{Z_{i1}Z_{i2}} = 0.650$	
$\gamma_5 = 0$	$c_5 = 0$	$\rho_{Z_{i1}Z_{i3}} = 0.700$	
$\gamma_6 = 0$	$c_6 = 0$	$\rho_{Z_{i2}Z_{i3}} = 0.750$	

(a)

Cumulants	Coefficients	Correlations
$\gamma_1 = 0$	$c_1 = -0.307740$	$\rho_{Y_{i1}Y_{i2}} = 0.650$
$\gamma_2 = 1$	$c_2 = 0.800560$	$\rho_{Y_{i1}Y_{i3}} = 0.700$
$\gamma_3 = 2$	$c_3 = 0.318764$	$\rho_{Y_{i2}Y_{i3}} = 0.750$
$\gamma_4 = 6$	$c_4 = 0.033500$	$\rho_{Z_{i1}Z_{i2}} = 0.690454$
$\gamma_5 = 24$	$c_5 = -0.003675$	$\rho_{Z_{i1}Z_{i3}} = 0.736781$
$\gamma_6 = 120$	$c_6 = 0.000159$	$\rho_{Z_{i2}Z_{i3}} = 0.782357$

(b)

Cumulants	Coefficients	Correlations
$\gamma_1 = 0$	$c_1 = 0$	$\rho_{Y_{i1}Y_{i2}} = 0.650$
$\gamma_2 = 1$	$c_2 = 1.248343$	$\rho_{Y_{i1}Y_{i3}} = 0.700$
$\gamma_3 = 0$	$c_3 = 0$	$\rho_{Y_{i2}Y_{i3}} = 0.750$
$\gamma_4 = -1$	$c_4 = -0.111426$	$\rho_{Z_{i1}Z_{i2}} = 0.660393$
$\gamma_5 = 0$	$c_5 = 0$	$\rho_{Z_{i1}Z_{i3}} = 0.709895$
$\gamma_6 = 48/7$	$c_6 = 0.004833$	$\rho_{Z_{i2}Z_{i3}} = 0.759104$

(c)

Cumulants	Coefficients	Correlations
$\gamma_1 = 0$	$c_1 = 0$	$\rho_{Y_{i1}Y_{i2}} = 0.650$
$\gamma_2 = 1$	$c_2 = 0.374011$	$\rho_{Y_{i1}Y_{i3}} = 0.700$
$\gamma_3 = 0$	$c_3 = 0$	$\rho_{Y_{i2}Y_{i3}} = 0.750$
$\gamma_4 = 25$	$c_4 = 0.159040$	$\rho_{Z_{i1}Z_{i2}} = 0.721687$
$\gamma_5 = 0$	$c_5 = 0$	$\rho_{Z_{i1}Z_{i3}} = 0.765685$
$\gamma_6 = 5000$	$c_6 = 0.002629$	$\rho_{Z_{i2}Z_{i3}} = 0.808143$

(d)

Note: Values of the intermediate correlations $\rho_{Z_{ii}Z_{ij}}$ are based on Y_{ii} and Y_{ij} having the same distribution in each class.

TABLE 4.21

Numerical Example to Demonstrate the Data Generation Process

1. Specify $k = 3$ classes where all three classes have one of the four (non)normal distributions depicted in Table 4.20. These distributions are described as: (A) Y_{i1} is standard normal, (B) Y_{i2} is asymmetric with a moderately heavy tail, (C) Y_{i3} is symmetric with light tails, (D)Y_{i4} is symmetric with heavy tails.
2. Specify $\rho_{I_2} = \bar{\rho} = 0.70 = \sum \rho_{Y_{ij}Y_{ij}}/k = (0.65 + 0.70 + 0.75)/3$. The pairwise correlations between $\rho_{Y_{ij}Y_{ij}}$ are listed in Table 4.20.
3. Determine the intermediate correlations $(\rho_{Z_{ij}Z_{ij}})$ by solving Equation 2.59 for each of the pairwise correlations. The intermediate correlations are listed in Table 4.20.
4. Specify $\rho_{I_1} = 0.60$, and thus using Equation 4.15 the variance component is $\sigma_\tau^2 = 0.1\bar{6}$.
5. Given $\rho_{I_1} = 0.60$, $\rho_{I_2} = 0.70$, and $\sigma_\tau^2 = 0.1\bar{6}$, a set of shift parameters that can be added to Y_{i1}, Y_{i2}, and Y_{i3} are $\mu_1 = -0.408248$, $\mu_2 = 0.0$, and $\mu_3 = 0.408248$, where
   $\sigma_\tau^2 = \sum \mu_i^2/(k-1) = [(-0.408248)^2 + (0)^2 + (0.408248)^2]/2 = 0.1\bar{6}$.
6. Given $k = 3$, $\rho_{I_2} = 0.70$, and $\sigma_\tau^2 = 0.1\bar{6}$, the other ICCs based on the formulae in Table 4.19 are $\rho_{I_3} = 0.8181\ldots$ and $\rho_{I_4} = 0.875$.
7. Assemble the intermediate correlations $(\rho_{Z_{ij}Z_{ij}})$ listed in Table 4.20 into 3×3 matrices and factor these matrices, e.g., using the *Mathematica* function CholeskyDecomposition. Use these results to generate standard normal deviates correlated at the intermediate levels. See Equation 2.56.
8. Use $k = 3$ polynomials of the form in Equation 4.17 with the coefficients in Table 4.20, values of μ_j from Step 5, and the standard normal deviates from Step 7 to generate the (non)normal distributions Y_{ij}.

For any given distribution and sample size, the 25,000 values of z^* were subsequently used to generate 25,000 bootstrap samples to obtain a 95% confidence interval on the mean of z^* using S-Plus (2007). The point estimates $\hat{\rho}_{I_t}$ and the lower and upper limits of the confidence intervals were then converted back to the original ICC metric and rounded to three decimal places. The results for the cases when the distributions were the same for each of the $k = 3$ classes are reported in Tables 4.22–4.25.

TABLE 4.22

ICC Estimates $\hat{\rho}_{I_t}$ of the Population Parameters $\rho_{I_{t=1,2,3,4}}$

	Sampling is from Distribution A in Table 4.21 with $k = 3$ classes and n objects in Figure 4.1.			
	$n = 10$	$n = 30$	$n = 100$	$n = 1000$
Population ICC	$\hat{\rho}_{I_t}$	$\hat{\rho}_{I_t}$	$\hat{\rho}_{I_t}$	$\hat{\rho}_{I_t}$
$\rho_{I_1} = .60$	.588 (.586, .590)	.597 (.596, .598)	.599 (.598, .599)	.600 (.600, .600)
$\rho_{I_2} = .70$	.690 (.688, .691)	.697 (.696, .698)	.699 (.698, .699)	.700 (.700, .700)
$\rho_{I_3} = .8181\ldots$	.805 (.804, .806)	.815 (.814, .816)	.817 (.817, .817)	.818 (.818, .818)
$\rho_{I_4} = .875$	.866 (.865, .867)	.873 (.872, .873)	.874 (.874, .874)	.875 (.875, .875)

Note: Each cell contains a 95% bootstrap confidence interval enclosed in parentheses.

TABLE 4.23

ICC Estimates $\hat{\rho}_{I_t}$ of the Population Parameters $\rho_{I_{t=1,2,3,4}}$

	Sampling is from Distribution B in Table 4.21 with $k = 3$ classes and n objects in Figure 4.1.			
	$n = 10$	$n = 30$	$n = 100$	$n = 1000$
Population ICC	$\hat{\rho}_{I_t}$	$\hat{\rho}_{I_t}$	$\hat{\rho}_{I_t}$	$\hat{\rho}_{I_t}$
$\rho_{I_1} = .60$	.562 (.560, .564)	.584 (.583, .585)	.595 (.594, .596)	.600 (.599, .600)
$\rho_{I_2} = .70$	.680 (.678, .682)	.690 (.689, .691)	.697 (.697, .698)	.700 (.700, .700)
$\rho_{I_3} = .8181...$	.785 (.783, .786)	.805 (.804, .806)	.814 (.814, .815)	.818 (.818, .818)
$\rho_{I_4} = .875$	.859 (.858, .860)	.868 (.867, .869)	.873 (.873, .874)	.875 (.875, .875)

Note: Each cell contains a 95% bootstrap confidence interval enclosed in parentheses.

TABLE 4.24

ICC Estimates $\hat{\rho}_{I_t}$ of the Population Parameters $\rho_{I_{t=1,2,3,4}}$

	Sampling is from Distribution C in Table 4.21 with $k = 3$ classes and n objects in Figure 4.1.			
	$n = 10$	$n = 30$	$n = 100$	$n = 1000$
Population ICC	$\hat{\rho}_{I_t}$	$\hat{\rho}_{I_t}$	$\hat{\rho}_{I_t}$	$\hat{\rho}_{I_t}$
$\rho_{I_1} = .60$	.599 (.597, .601)	.601 (.600, .602)	.600 (.599, .600)	.600 (.600, .600)
$\rho_{I_2} = .70$	.701 (.699, .703)	.701 (.700, .702)	.700 (.700, .701)	.700 (.700, .700)
$\rho_{I_3} = .8181...$	.812 (.811, .813)	.817 (.816, .818)	.818 (.817, .818)	.818 (.818, .818)
$\rho_{I_4} = .875$	.872 (.871, .873)	.874 (.873, .875)	.875 (.874, .875)	.875 (.875, .875)

Note: Each cell contains a 95% bootstrap confidence interval enclosed in parentheses.

TABLE 4.25

ICC Estimates $\hat{\rho}_{I_t}$ of the Population Parameters $\rho_{I_{t=1,2,3,4}}$

	Sampling is from Distribution D in Table 4.21 with $k = 3$ classes and n objects in Figure 4.1.			
	$n = 10$	$n = 30$	$n = 100$	$n = 1000$
Population ICC	$\hat{\rho}_{I_t}$	$\hat{\rho}_{I_t}$	$\hat{\rho}_{I_t}$	$\hat{\rho}_{I_t}$
$\rho_{I_1} = .60$	.537 (.535, .539)	.572 (.570, .573)	.589 (.588, .590)	.599 (.599, .599)
$\rho_{I_2} = .70$	.677 (.675, .679)	.691 (.690, .692)	.697 (.696, .698)	.700 (.700, .700)
$\rho_{I_3} = .8181...$	.767 (.765, .769)	.797 (.796, .798)	.810 (.810, .811)	.817 (.817, .818)
$\rho_{I_4} = .875$	.859 (.857, .860)	.869 (.868, .869)	.873 (.872, .873)	.875 (.875, .875)

Note: Each cell contains a 95% bootstrap confidence interval enclosed in parentheses.

TABLE 4.26

ICC Estimates $\hat{\rho}_{I_t}$ of the Population Parameters $\rho_{I_{t=1,2,3,4}}$

	Sampling is from Distribution A, B, or C in Table 4.21 for Class 1, 2, or 3 with n objects in Figure 4.1.			
	A-1, B-2, C-3	**A-1, B-2, C-3**	**A-1, B-2, C-3**	**A-1, B-2, C-3**
	$n = 10$	$n = 30$	$n = 100$	$n = 1000$
Population ICC	$\hat{\rho}_{I_t}$	$\hat{\rho}_{I_t}$	$\hat{\rho}_{I_t}$	$\hat{\rho}_{I_t}$
$\rho_{I_1} = .60$	.590 (.588, .591)	.598 (.597, .599)	.599 (.599, .600)	.600 (.600, .600)
$\rho_{I_2} = .70$	.694 (.692, .695)	.700 (.699, .700)	.700 (.700, .700)	.700 (.700, .700)
$\rho_{I_3} = .8181...$	.807 (.806, .808)	.815 (.815, .816)	.817 (.817, .818)	.818 (.818, .818)
$\rho_{I_4} = .875$	.869 (.868, .870)	.874 (.874, .875)	.875 (.875, .875)	.875 (.875, .875)

Note: Each cell contains a 95% bootstrap confidence interval enclosed in parentheses.

The methodology described above can also be applied to cases when the distributions vary across the $k = 3$ classes. Specifically, let us consider classes 1, 2, and 3 to have distributions A, B, and C as given in Table 4.21, respectively. All that is necessary is to obtain the intermediate correlations for these distributions. As such, the intermediate correlations are $\rho_{Z_{i1}Z_{i2}} = 0.719471$, $\rho_{Z_{i1}Z_{i3}} = 0.709536$, and $\rho_{Z_{i2}Z_{i3}} = 0.850594$. The simulation results for this scenario are reported in Table 4.26.

To demonstrate empirically that the methodology simulates consistent estimates, single samples of size $n = 1,000,000$ were also drawn from the four distributions in Table 4.21. The same four ICC statistics were computed on these data using the formulae in Table 4.18 but without the Hotelling (1953) z^* transformation. These results are reported in Table 4.27 for cases where the distributions were either the same or different across the three classes.

TABLE 4.27

ICC Estimates $\hat{\rho}_{I_t}$ of the Parameters $\rho_{I_{t=1,2,3,4}}$

Population ICC	A-1, A-2, A-3 $\hat{\rho}_{I_t}$	B-1, B-2, B-3 $\hat{\rho}_{I_t}$	C-1, C-2, C-3 $\hat{\rho}_{I_t}$	D-1, D-2, D-3 $\hat{\rho}_{I_t}$	A-1, B-2, C-3 $\hat{\rho}_{I_t}$
$\rho_{I_1} = .60$	.600	.600	.600	.600	.600
$\rho_{I_2} = .70$	.700	.700	.700	.700	.700
$\rho_{I_3} = .8181...$	.818	.818	.818	.818	.818
$\rho_{I_4} = .875$	.875	.875	.875	.875	.875

Note: Sampling is from Distribution A, B, C, or D in Table 4.21 in Class 1, 2, or 3 for $n = 10^6$ objects in Figure 4.1.

The simulation results reported in Tables 4.22–4.26 indicate that the procedure generates empirical estimates that were in close agreement with their associated population parameters as the sample sizes increased. This was demonstrated whether the classes had the same or different distributions. Further, the standard normal distribution and the symmetric distribution with light tails generated ICC estimates of the population parameters that were in close agreement even for sample sizes as small as $n = 10$. However, the asymmetric distribution and the symmetric distribution with heavy tails required larger samples sizes (e.g., $n = 100$) for some of the ICC estimates to approach their respective population parameters.

Inspection of Table 4.27 indicates that the procedure generates consistent ICC estimates of the population parameters. Specifically, in all cases, each estimate is the same as its associated population parameter when rounding is to three decimal places. It is also worthy to note that the method described in this section is also appropriate for standard uniform- or logistic-based polynomials, as was demonstrated in the previous section for systems of linear statistical models.

4.9 Simulating Correlated Continuous Variates and Ranks

In this section we turn the focus of our attention to simulating correlated nonnormal continuous distributions with the rank orders of variate values associated power method polynomials. Let us first consider the variates Y_i and Y_j to be fifth-order normal-based power method polynomials as

$$Y_i = \sum_{m=1}^{6} c_{mi} Z_i^{m-1} \tag{4.19}$$

$$Y_j = \sum_{m=1}^{6} c_{mj} Z_j^{m-1} \tag{4.20}$$

Depicted in Figure 4.2 is a schematic of the bivariate correlation structure for the polynomials associated with Equations 4.19, 4.20, and their respective rank orders, $R(Y_i)$ and $R(Y_j)$.

Having the ability to control the correlation structure in Figure 4.2 would allow one, for example, to simulate correlation structures between continuous nonnormal variates and ranks. Specifically, this would require us to determine the intermediate correlation between Z_i and Z_j such that Y_i and $R(Y_j)$ would have a user-specified correlation $\rho_{Y_i, R(Y_j)}$. Some applications where this could be useful would be simulating correlation structures between (1) percentile (or class) ranks and achievement test scores, (2) variate values and

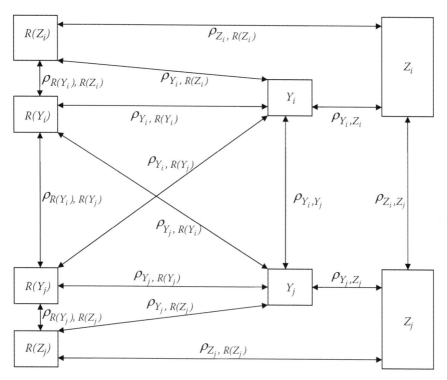

FIGURE 4.2
Schematic of the bivariate correlation structure for the power method based on Equations (4.19) and (4.20) and their rank orders.

ranked data in the context of multivariate statistics (e.g., Hotelling's T^2 where one variable is continuous and the other is a rank), or (3) rank-transformed covariates and a nonnormal variate $\rho_{Y_i,R(Y_j)}$ or its rank $\rho_{R(Y_i),R(Y_j)}$ in the context of the general linear model.

Moreover, it would also be useful to have a measure of the amount of potential loss of information when using ranks $R(Y_i)$ in place of the variate values Y_i. One approach that could be used to address this concern would be analogous to Stuart's (1954) use of the product-moment coefficient of correlation between the variate values and their rank order (i.e., $\rho_{Y_i,R(Y_i)}$ in Figure 4.2) as the measure of this potential loss in efficiency. This measure is based on the coefficient of mean difference denoted as Δ (Gini, 1912). More specifically, if two independent continuous variables X_1 and X_2 are both distributed as X, then Δ can be expressed as (Johnson, Kotz, & Balakrishnan, 1994, p. 3; Kendall & Stuart, 1977, p. 47)

$$\Delta = E[|X_1 - X_2|] = \int_{-\infty}^{+\infty} \int_{-\infty}^{+\infty} |x_1 - x_2|\, f(x_1) f(x_2)\, dx_1\, dx_2 \qquad (4.21)$$

It follows from Stuart (1954) that the correlation between the variate X and its rank order $R(X)$ is

$$\rho_{X,R(X)} = \sqrt{(n-1)/(n+1)}(\Delta\sqrt{3})/(2\sigma) \tag{4.22}$$

where σ is the population standard deviation associated with X. An alternative derivation of Equation 4.22 is also given in Gibbons and Chakraborti (1992, pp. 132–135). Some specific evaluations of Equation 4.22 for various well-known pdfs (e.g., gamma, normal, uniform) are given in Stuart (1954) or in Kendall and Gibbons (1990, pp. 165–166). The methodology and formulae for the correlation structure in Figure 4.2 are subsequently presented.

4.10 Methodology

We assume that the variates Y_i and Y_j in Equation 4.19 and Equation 4.20 produce valid power method pdfs and are thus increasing monotonic transformations in Z_i and Z_j. This implies that the rank orders of Y_i (Y_j) and Z_i (Z_j) are identical and thus will have rank correlations of $\rho_{R(Y_i),R(Z_i)} = \rho_{R(Y_j),R(Z_j)} = 1$ in Figure 4.2.

Given these assumptions, let Z_i and Z_j have univariate and bivariate pdfs express as

$$f_i := f_{Z_i}(z_i) = (2\pi)^{-\frac{1}{2}}\exp\{-z_i^2/2\} \tag{4.23}$$

$$f_j := f_{Z_j}(z_j) = (2\pi)^{-\frac{1}{2}}\exp\{-z_j^2/2\} \tag{4.24}$$

$$f_{ij} := f_{Z_iZ_j}(z_i, z_j, \rho_{z_iz_j}) = \left(2\pi\sqrt{1-\rho_{z_iz_j}^2}\right)^{-1}\exp\left\{-\left(2\left(1-\rho_{z_iz_j}^2\right)\right)^{-1}\left(z_i^2 - 2\rho_{z_iz_j}z_iz_j + z_j^2\right)\right\} \tag{4.25}$$

with distribution functions $\Phi(z_i)$ and $\Phi(z_j)$ as in Equation 2.50 and Equation 2.51. As such, $\Phi(z_i)$ and $\Phi(z_j)$ will be uniformly distributed, that is, $\Phi(z_i) \sim U_i[0,1]$, $\Phi(z_j) \sim U_j[0,1]$, with correlation $\rho_{\Phi(z_i),\Phi(z_j)} = (6/\pi)\sin^{-1}(\rho_{z_iz_j}/2)$ (Pearson, 1907).

If Z_i and Z_j are independent, then from Equations 4.21–4.24 the coefficient of mean difference for the unit normal distribution is

$$\Delta = 2/\sqrt{\pi} = E[|Z_i - Z_j|] = \int_{-\infty}^{+\infty}\int_{-\infty}^{+\infty}|z_i - z_j|\,f_if_j\,dz_i\,dz_j \tag{4.26}$$

Substituting Δ from Equation 4.26 into Equation 4.22 yields for the standard normal distribution

$$\rho_{Z_i,R(Z_i)} = \sqrt{(n-1)/(n+1)}\sqrt{3/\pi} \tag{4.27}$$

Using Equation 4.19 and Equation 4.20 with $c_{mi} = c_{mj}$, and Equation 4.26, the coefficient of mean difference for the power method is expressed as

$$\Delta = E\left[\left|\sum_{m=1}^{6} c_{mi}Z_i^{m-1} - \sum_{m=1}^{6} c_{mj}Z_j^{m-1}\right|\right]$$

$$= \int_{-\infty}^{+\infty}\int_{-\infty}^{+\infty}\left|\sum_{m=1}^{6} c_{mi}z_i^{m-1} - \sum_{m=1}^{6} c_{mj}z_j^{m-1}\right| f_i f_j \, dz_i \, dz_j \tag{4.28}$$

Changing from rectangular to polar coordinates in Equation 4.28, to eliminate the absolute value, and integrating yields

$$\Delta = \int_{0}^{\frac{\pi}{4}}\int_{0}^{+\infty}\left(\sum_{m=1}^{6} c_{mi}(r\cos\theta)^{m-1} - \sum_{m=1}^{6} c_{mi}(r\sin\theta)^{m-1}\right)\frac{r}{2\pi}\exp\{-r^2/2\}\,dr\,d\theta$$

$$- \int_{\frac{\pi}{4}}^{\frac{5\pi}{4}}\int_{0}^{+\infty}\left(\sum_{m=1}^{6} c_{mi}(r\cos\theta)^{m-1} - \sum_{m=1}^{6} c_{mi}(r\sin\theta)^{m-1}\right)\frac{r}{2\pi}\exp\{-r^2/2\}\,dr\,d\theta$$

$$+ \int_{\frac{5\pi}{4}}^{2\pi}\int_{0}^{+\infty}\left(\sum_{m=1}^{6} c_{mi}(r\cos\theta)^{m-1} - \sum_{m=1}^{6} c_{mi}(r\sin\theta)^{m-1}\right)\frac{r}{2\pi}\exp\{-r^2/2\}\,dr\,d\theta$$

$$\Delta = (2\sqrt{\pi})^{-1}(4c_{2i} + 10c_{4i} + 43c_{6i}) \tag{4.29}$$

Substituting Δ from Equation 4.29 into Equation 4.22 and setting $\sigma = 1$, because all power method distributions have unit variances, gives the correlations $\rho_{Y_i,R(Y_i)}$ and $\rho_{Y_i,R(Z_i)}$ in Figure 4.2 as

$$\rho_{Y_i,R(Y_i)} = \rho_{Y_i,R(Z_i)} = \sqrt{(n-1)/(n+1)}\sqrt{3/\pi}((1/4)(4c_{2i} + 10c_{4i} + 43c_{6i})) \tag{4.30}$$

Thus, Equation 4.30 is a normal-based polynomial's measure of the potential loss of information when using ranks in place of variate values. If we were to consider, for example, substituting the even subscripted coefficients ($c_{2,4,6}$) associated with the fifth-order polynomial in Table 3.4 into Equation 4.30, then we would obtain as $n \to \infty$, $\rho_{Y_i,R(Y_i)} \approx 0.900283$ which is close to the exact correlation for the chi-square distribution with $df = 3$ based on Equation 4.22 of $\rho_{X,R(X)} = 2\sqrt{2}/\pi \approx 0.900316$, where $\Delta = 8/\pi$ and $\sigma = \sqrt{6}$. Note also that for the special case where $Y_i = Z_i$ in Equation 4.19, that is, $c_{2i} = 1$ and $c_{(m \neq 2)i} = 0$, then Equation 4.30 reduces to Equation 4.27.

Remark 3.1

The strictly increasing monotone assumption in Equation 4.19, that is, $Y_i' > 0$, implies that $\rho_{Y_i,Z_i} \in (0,1]$. Thus, for Equations 4.29 and 4.30 to both be positive, the restriction on the constants c_{2i}, c_{4i}, and c_{6i} is that they must satisfy the inequality $0 < c_{2i} + 3c_{4i} + 15c_{6i} \leq 1$.

Proof

The correlation ρ_{Y_i,Z_i} (see Figure 4.2) can be determined by equating $\rho_{Y_i,Z_i} = E[Y_i Z_i]$ because both Y_i and Z_i have zero means and unit variances. Hence,

$$\rho_{Y_i,Z_i} = E[Y_i Z_i] = E\left[\sum_{m=1}^{6} c_{mi} Z_i^m\right] = \sum_{m=1}^{6} c_{mi} E[Z_i^m] = c_{2i} + 3c_{4i} + 15c_{6i} \qquad (4.31)$$

because all odd central moments of the standard normal distribution are zero and the even central moments are $\mu_2 = 1, \mu_4 = 3$, and $\mu_6 = 15$.

The correlations ρ_{Y_i,Y_j} and $\rho_{Y_i,R(Y_j)}$ in Figure 4.2 can be obtained by making use of Equations 4.19, 4.20, 4.25, and the standardized form of Equation 2.50. More specifically, ρ_{Y_i,Y_j} and $\rho_{Y_i,R(Y_j)}$ can be numerically calculated from

$$\rho_{Y_i,Y_j} = \int_{-\infty}^{+\infty} \int_{-\infty}^{+\infty} \left(\sum_{m=1}^{6} c_{mi} z_i^{m-1}\right)\left(\sum_{m=1}^{6} c_{mj} z_j^{m-1}\right) f_{ij} dz_i dz_j \qquad (4.32)$$

$$\rho_{Y_i,R(Y_j)} = \rho_{Y_i,R(Z_j)}$$

$$= \sqrt{(n-1)/(n+1)} \int_{-\infty}^{+\infty} \int_{-\infty}^{+\infty} \left(\sum_{m=1}^{6} c_{mi} z_i^{m-1}\right)(\sqrt{3}(2\Phi(z_j)-1)) f_{ij} dz_i dz_j \qquad (4.33)$$

because $\rho_{R(Z_j),\Phi(Z_j)} = \rho_{\Phi(Z_j),R(\Phi(Z_j))} = \sqrt{(n-1)/(n+1)}$ since $\Phi(z_j) \sim U_j[0,1]$, and in (4.22) $\Delta = 1/3$ and $\sigma = 1/\sqrt{12}$ for the regular uniform distribution. We note that (4.32) is Equation 2.55 for the special case of the normal distribution and is also algebraically equivalent to Equation 2.59.

The correlation $\rho_{R(Y_i),R(Y_j)}$ in Figure 4.2 can be obtained from the derivation of $\rho_{R(Z_i),R(Z_j)}$ given in Moran (1948). That is, because Equations 4.19 and 4.20 are strictly increasing monotonic transformations, $\rho_{R(Y_i),R(Y_j)} = \rho_{R(Z_i),R(Z_j)}$ and thus

$$\rho_{R(Y_i),R(Y_j)} = (6/\pi)\left[((n-2)/(n+1))\sin^{-1}(\rho_{Z_i,Z_j}/2) + (1/(n+1))\sin^{-1}(\rho_{Z_i,Z_j})\right] \qquad (4.34)$$

TABLE 4.28

Correlation Structure in Figure 4.2 for Fifth-Order Normal-Based Polynomials

(a)	$\rho_{R(Y_i),R(Z_i)} = 1$
(b)	$\rho_{Y_i,Z_i} = c_{2i} + 3c_{4i} + 15c_{6i}$
(c)	$\rho_{Z_i,R(Z_i)} = \sqrt{(n-1)/(n+1)}\sqrt{3/\pi}$
(d)	$\rho_{Y_i,Y_j} = \int_{-\infty}^{+\infty}\int_{-\infty}^{+\infty}\left(\sum_{m=1}^{6} c_{mi}z_i^{m-1}\right)\left(\sum_{m=1}^{6} c_{mj}z_j^{m-1}\right)f_{ij}dz_i dz_j$
(e)	$\rho_{Y_i,R(Y_i)} = \sqrt{(n-1)/(n+1)}\sqrt{3/\pi}((1/4)(4c_{2i} + 10c_{4i} + 43c_{6i}))$
(f)	$\rho_{Y_i,R(Y_j)} = \sqrt{(n-1)/(n+1)}\int_{-\infty}^{+\infty}\int_{-\infty}^{+\infty}\left(\sum_{m=1}^{6} c_{mi}z_i^{m-1}\right)(\sqrt{3}(2\Phi(z_j)-1))f_{ij}dz_i dz_j$
(g)	$\rho_{R(Y_i),R(Y_j)} = (6/\pi)[((n-2)/(n+1))\sin^{-1}(\rho_{Z_i,Z_j}/2) + (1/(n+1))\sin^{-1}(\rho_{Z_i,Z_j})]$

Further, note that as $n \to \infty$, (4.34) is algebraically equivalent to

$$(6/\pi)\sin^{-1}(\rho_{Z_iZ_j}/2) = \int_{-\infty}^{+\infty}\int_{-\infty}^{+\infty}(\sqrt{3}(2\Phi(z_i)-1))(\sqrt{3}(2\Phi(z_j)-1))f_{ij}dz_i dz_j \quad (4.35)$$

where the left-hand side of Equation 4.35 corresponds with the expression derived by Pearson (1907). Table 4.28 gives a summary of the correlation structure in Figure 4.2.

4.11 Numerical Example and Monte Carlo Simulation

To evaluate the methodology in the previous section, four (non)normal distributions ($Y_{i=1,2,3,4}$) were simulated with the specified standardized cumulants (γ_i) and correlations ($\rho_{Y_i,Z_i}, \rho_{Y_i,R(Y_i)}$) listed in Table 4.29. The correlations ρ_{Y_i,Z_i} and $\rho_{Y_i,R(Y_i)}$ were determined from using Equations (b) and (e) in Table 4.28.

The specified correlations of concern for this simulation are listed in Table 4.30. These parameters represent examples of correlations between variates (ρ_{Y_i,Y_j}), variates with ranks ($\rho_{Y_i,R(Y_j)}$), and ranks ($\rho_{R(Y_i),R(Y_j)}$). Tables 4.31–4.33 give the required intermediate correlations for simulating the specified correlations in Table 4.30 for sample sizes of $n = 10, n = 30$, and $n = 10^6$.

The intermediate correlations associated with the specified correlations of ρ_{Y_i,Y_j}, $\rho_{Y_i,R(Y_j)}$, and $\rho_{R(Y_i),R(Y_j)}$ were determined from Equations (d), (f), and (g) in Table 4.28. Presented in Table 4.34 is an example of solving for the intermediate correlation associated with $\rho_{Y_2,R(Y_4)}$ in Table 4.30 and Equation (f) in Table 4.28 using the *Mathematica* source code.

The data generation procedure began by conducting Cholesky factorizations on each of the intermediate correlation matrices in Tables 4.31–4.33.

TABLE 4.29

Fifth-Order Normal-Based Power Method Distributions for the Simulation

Cumulants	Constants	Correlations	Probability Density Function for $Y_{i=1,2,3,4}$
$\gamma_1 = 0$	$c_{11} = 0$	$\rho_{Y_1,Z_1} = 1.0$	
$\gamma_2 = 1$	$c_{12} = 1$	$\rho_{Y_1,R(Y_1),n=10} = .883915$	
$\gamma_3 = 0$	$c_{13} = 0$	$\rho_{Y_1,R(Y_1),n=30} = .945156$	
$\gamma_4 = 0$	$c_{14} = 0$	$\rho_{Y_1,R(Y_1),n=10^6} = .977204$	
$\gamma_5 = 0$	$c_{15} = 0$		
$\gamma_6 = 0$	$c_{16} = 0$		

$\gamma_1 = 0$	$c_{21} = -0.307740$	$\rho_{Y_2,Z_2} = .903441$	
$\gamma_2 = 1$	$c_{22} = 0.800560$	$\rho_{Y_2,R(Y_2),n=10} = .783164$	
$\gamma_3 = 2$	$c_{23} = 0.318764$	$\rho_{Y_2,R(Y_2),n=30} = .837425$	
$\gamma_4 = 6$	$c_{24} = 0.033500$	$\rho_{Y_2,R(Y_2),n=10^6} = .865819$	
$\gamma_5 = 24$	$c_{25} = -0.003675$		
$\gamma_6 = 120$	$c_{26} = 0.000159$		

$\gamma_1 = 0$	$c_{31} = 0$	$\rho_{Y_3,Z_3} = .890560$	
$\gamma_2 = 1$	$c_{32} = 0.374011$	$\rho_{Y_3,R(Y_3),n=10} = .707015$	
$\gamma_3 = 0$	$c_{33} = 0$	$\rho_{Y_3,R(Y_3),n=30} = .756001$	
$\gamma_4 = 25$	$c_{34} = 0.159040$	$\rho_{Y_3,R(Y_3),n=10^6} = .781634$	
$\gamma_5 = 0$	$c_{35} = 0$		
$\gamma_6 = 5000$	$c_{36} = 0.002629$		

$\gamma_1 = 0$	$c_{41} = 0$	$\rho_{Y_4,Z_4} = .986561$	
$\gamma_2 = 1$	$c_{42} = 1.248343$	$\rho_{Y_4,R(Y_4),n=10} = .903126$	
$\gamma_3 = 0$	$c_{43} = 0$	$\rho_{Y_4,R(Y_4),n=30} = .965699$	
$\gamma_4 = -1$	$c_{44} = -0.111426$	$\rho_{Y_4,R(Y_4),n=10^6} = .998442$	
$\gamma_5 = 0$	$c_{45} = 0$		
$\gamma_6 = 48/7$	$c_{46} = 0.004833$		

TABLE 4.30

Specified Correlations for the Distributions in Table 4.29

$\rho_{Y_1,Y_2} = 0.70$		
$\rho_{Y_1,Y_3} = 0.60$	$\rho_{Y_2,R(Y_3)} = 0.60$	
$\rho_{Y_1,Y_4} = 0.40$	$\rho_{Y_2,R(Y_4)} = 0.50$	$\rho_{R(Y_3),R(Y_4)} = 0.80$

TABLE 4.31

Intermediate Correlation Matrix for the Correlations in Table 4.30 with $n = 10$

	Z_1	Z_2	Z_3	Z_4
Z_1	1			
Z_2	0.774815	1		
Z_3	0.664127	0.759832	1	
Z_4	0.442751	0.630988	0.862571	1

TABLE 4.32

Intermediate Correlation Matrix for the Correlations in Table 4.30 with $n = 30$

	Z_1	Z_2	Z_3	Z_4
Z_1	1			
Z_2	0.774815	1		
Z_3	0.664127	0.709578	1	
Z_4	0.442751	0.589522	0.831382	1

TABLE 4.33

Intermediate Correlation Matrix for the Correlations in Table 3 with $n = 10^6$

	Z_1	Z_2	Z_3	Z_4
Z_1	1			
Z_2	0.774815	1		
Z_3	0.664127	0.685865	1	
Z_4	0.442751	0.569937	0.813474	1

TABLE 4.34

Mathematica Code for Estimating the Intermediate Correlation $\hat{\rho}_{z_2 z_4}$ in Table 4.31

(* Specified correlation in Table 4.30: $\rho_{Y_2, R(Y_4)} = 0.50$

The estimation of $\hat{\rho}_{z_2 z_4}$ requires the use of equation (f) in Table 4.28

Estimated intermediate correlation listed in Table 4.31 (see below): $\hat{\rho}_{z_2 z_4} = 0.630988$ *)

$c_{21} = -0.3077396;$

$c_{22} = 0.8005604;$

$c_{23} = 0.3187640;$

$c_{24} = 0.0335001;$

$c_{25} = -0.0036748;$

$c_{26} = 0.0001587;$

$n = 10;$

$\hat{\rho}_{z_2 z_4} = 0.630988;$

$$\Phi_4 = \int_{-\infty}^{z_4} (\sqrt{2\pi})^{-1} e^{-\frac{u_4^2}{2}} du_4;$$

$$Y_2 = \sum_{k=1}^{6} c_{2k} z_2^{k-1};$$

$$f_{24} = \left(2 * \pi * \left(1 - \rho_{z_2 z_4}^2\right)^{\frac{1}{2}}\right)^{-1} * \mathrm{Exp}\left[-\left(2 * \left(1 - \rho_{z_2 z_4}^2\right)\right)^{-1} * \left(z_2^2 - 2 * \rho_{z_2 z_4} * z_2 * z_4 + z_4^2\right)\right];$$

$$\mathrm{int} = \sqrt{\frac{n-1}{n+1}} \mathrm{NIntegrate}[(Y_2(\sqrt{3}(2\Phi_4 - 1))) f_{24}, \{z_2, -6, 6\}, \{z_4, -6, 6\}, \mathrm{Method} \to \mathrm{Trapezoidal}]$$

Solution: $\mathrm{int} = 0.49999987969495174$

The entries from the factored matrices were then used to produce standard normal deviates with intercorrelations equal to the intermediate correlations as described in Section 2.6 and using Equation 2.56. These deviates were subsequently transformed by using the constants c_{mi} in Table 4.29 and the polynomials of the form in Equations 4.19 and 4.20 to produce the specified (non)normal distributions with the specified correlations in Table 4.30 for each of the three different sample sizes.

Overall average estimates of the cumulants ($\hat{\gamma}_1 = \mathrm{mean}, \hat{\gamma}_2 = \mathrm{variance}, \hat{\gamma}_3 = $ skew, and $\hat{\gamma}_4 = \mathrm{kurtosis}$) and correlations ($\hat{\rho}_{Y_i, R(Y_i)}$, $\hat{\rho}_{Y_i Y_j}$, $\hat{\rho}_{Y_i R(Y_j)}$, and $\hat{\rho}_{R(Y_i)R(Y_j)}$) were obtained and were based on $(n = 10) \times 100,000$ and $(n = 30) \times 100,000$ pseudorandom deviates. Estimates of the standardized cumulants and correlations were also obtained based on single draws of size $n = 10^6$.

The empirical estimates of $\hat{\gamma}_i$ and $\hat{\rho}_{Y_i, R(Y_i)}$ are reported in Tables 4.35–4.37. The estimates of $\hat{\rho}_{Y_i Y_j}$, $\hat{\rho}_{Y_i R(Y_j)}$, and $\hat{\rho}_{R(Y_i)R(Y_j)}$ are listed in Tables 4.38–4.40. Inspection of the results presented in Tables 4.35–4.40 indicates that all estimates were in close proximity with their specified parameter even for sample sizes as small as $n = 10$.

TABLE 4.35

Empirical Estimates of the First Four Standardized Cumulants ($\hat{\gamma}_i$) and the Variate Value and Rank Correlation $\hat{\rho}_{Y_i, R(Y_i)}$ Listed in Table 4.29

Distribution	Mean ($\hat{\gamma}_1$)	Variance ($\hat{\gamma}_2$)	Skew ($\hat{\gamma}_3$)	Kurtosis ($\hat{\gamma}_4$)	$\hat{\rho}_{Y_i, R(Y_i)}$
Y_1	0.000	1.005	0.000	0.002	.884
Y_2	0.001	1.002	2.007	6.021	.783
Y_3	0.000	1.001	0.000	25.181	.707
Y_4	0.000	1.000	0.001	−1.003	.903

Note: Sample size is $n = 10$.

TABLE 4.36

Empirical Estimates of the First Four Standardized Cumulants ($\hat{\gamma}_i$) and the Variate Value and Rank Correlation $\hat{\rho}_{Y_i, R(Y_i)}$ Listed in Table 4.29

Distribution	Mean ($\hat{\gamma}_1$)	Variance ($\hat{\gamma}_2$)	Skew ($\hat{\gamma}_3$)	Kurtosis ($\hat{\gamma}_4$)	$\hat{\rho}_{Y_i, R(Y_i)}$
Y_1	0.000	1.000	0.000	0.000	.945
Y_2	0.000	1.000	2.000	5.99	.837
Y_3	0.000	0.999	0.008	25.067	.756
Y_4	0.000	1.000	0.000	−1.003	.965

Note: Sample size is $n = 30$.

TABLE 4.37

Empirical Estimates of the First Four Standardized Cumulants ($\hat{\gamma}_i$) and the Variate Value and Rank Correlation $\hat{\rho}_{Y_i, R(Y_i)}$ Listed in Table 4.29

Distribution	Mean ($\hat{\gamma}_1$)	Variance ($\hat{\gamma}_2$)	Skew ($\hat{\gamma}_3$)	Kurtosis ($\hat{\gamma}_4$)	$\hat{\rho}_{Y_i, R(Y_i)}$
Y_1	0.000	0.997	0.000	−0.014	.977
Y_2	0.001	1.000	2.012	5.98	.866
Y_3	0.000	0.993	0.080	24.976	.782
Y_4	0.000	0.997	0.000	−1.015	.998

Note: Sample size is $n = 10^6$.

TABLE 4.38

Empirical Estimates of the Correlations in Table 4.30 for $n = 10$

$\hat{\rho}_{Y_1, Y_2} = 0.700$

$\hat{\rho}_{Y_1, Y_3} = 0.602$ $\qquad$ $\hat{\rho}_{Y_2, R(Y_3)} = 0.601$

$\hat{\rho}_{Y_1, Y_4} = 0.401$ $\qquad$ $\hat{\rho}_{Y_2, R(Y_4)} = 0.501$ $\qquad$ $\rho_{R(Y_3), R(Y_4)} = 0.800$

TABLE 4.39

Empirical Estimates of the Correlations in Table 4.30 for $n = 30$

$\hat{\rho}_{Y_1,Y_2} = 0.700$		
$\hat{\rho}_{Y_1,Y_3} = 0.600$	$\hat{\rho}_{Y_2,R(Y_3)} = 0.600$	
$\hat{\rho}_{Y_1,Y_4} = 0.400$	$\hat{\rho}_{Y_2,R(Y_4)} = 0.501$	$\hat{\rho}_{R(Y_3),R(Y_4)} = 0.800$

TABLE 4.40

Estimates of the Correlations in Table 4.30 Using a Single Draw of $n = 10^6$

$\hat{\rho}_{Y_1,Y_2} = 0.700$		
$\hat{\rho}_{Y_1,Y_3} = 0.602$	$\hat{\rho}_{Y_2,R(Y_3)} = 0.600$	
$\hat{\rho}_{Y_1,Y_4} = 0.401$	$\hat{\rho}_{Y_2,R(Y_4)} = 0.500$	$\hat{\rho}_{R(Y_3),R(Y_4)} = 0.800$

4.12 Some Additional Comments

As mentioned in Chapter 1, many methodologists have used power method polynomials to simulate nonnormal data for the purpose of examining Type I error and power properties among competing (non)parametric and other rank-based statistical procedures. As such, of particular importance is the correlation $\rho_{Y_i,R(Y_i)}$ in Table 4.28, which can be used as a measure for the potential loss of information when using the ranks in place of the variate values. For example, distribution Y_3 in Table 4.29 has a correlation between its variate values and assigned ranks that is rather low ($\rho_{Y_3,R(Y_3)},n=30 \approx .756$) relative to the other pdfs (e.g., Y_1 and Y_4). As such, conclusions from inferential procedures based on the ranks of Y_3 could differ markedly from analyses performed on the actual variate values. In contrast, distribution Y_4 has a variate rank correlation of $\rho_{Y_4,R(Y_4)},n=30 \approx .966$. In this case, with such a high correlation, conclusions from statistical analyses based on the ranks would usually not differ much from the same analyses performed on the variate values (Gibbons & Chakraborti, 1992, p. 132; Kendall & Gibbons, 1990, p. 166).

To demonstrate, a small Monte Carlo study was conducted to compare the relative Type I error and power properties of the variate values and their rank orders for a two-group independent samples design with $n_A = n_B = 30$. We used the two symmetric distributions Y_3 and Y_4 (in Table 4.29), which have the disparate variate rank correlations indicated above.

The data for Groups A and B were randomly sampled from either Y_3 or Y_4, where both groups were sampled from the same (symmetric) distribution. In each condition, the usual OLS parametric t-test was applied to both the variate values and their assigned ranks. It is noted that an OLS procedure applied to ranked data in this context is referred to as a rank-transform (RT) procedure suggested by Conover and Iman (1981), and is functionally equivalent to the Mann–Whitney–Wilcoxon test. Thus, with symmetric error distributions, the null hypothesis for the rank-based analysis $H_0 : P(Y_A < Y_B) = P(Y_A > Y_B)$ is equivalent to the null hypothesis for the parametric t-test $H_0 : \mu_A = \mu_B$; that is, $E[Y_A] = E[Y_B]$ is equivalent to $E[R(Y_A)] = E[R(Y_B)]$ (see Vargha & Delaney, 1998, 2000).

Empirical Type I error and power rates were based on 50,000 replications within each condition in the Monte Carlo study. In terms of Type I error (nominal $\alpha = 0.05$), when both groups were sampled from Y_3, the empirical Type I error rate for the parametric t-test performed on the variate values was conservative $\tilde{\alpha} = 0.040$, whereas the rejection rate for the t-test performed on the ranks was close to nominal alpha $\hat{\alpha} = 0.051$. In contrast, when both groups were sampled from Y_4, the empirical Type I errors rates were essentially the same (to three decimals places) for both tests, $\tilde{\alpha} = \hat{\alpha} = 0.051$.

In terms of power analysis, the rejection rates for the tests differed markedly when sampling was from Y_3. For example, with a standardized effect size of one-half a standard deviation between the variate means, the rejection rates were 0.55 for the parametric t-test performed on the variate values and 0.89 for the t-test performed on the ranks. In contradistinction, when sampling was from Y_4, and with the same standardized effect size of 0.50, the rejection rates were similar for the parametric t-test and the RT, 0.46 and 0.44, respectively.

Some other applications of the proposed methodology that are worth mentioning are in terms of the use of ranks in principal components analysis (PCA) or exploratory factor analysis (EFA) (Bartholomew, 1983; Besse & Ramsay, 1986). More specifically, this method not only controls the correlation structure between nonnormal variates but also allows the researcher insight into how ranking will affect the subsequent correlation matrix submitted to a PCA or EFA. Thus, knowing the correlation between nonnormal variates and their ranks is useful for investigating estimation bias when using rank-based statistics.

It is also worth noting that the methodology described in Section 4.10 is also applicable to uniform or logistic-based power method polynomials. However, there are some differences that need to be mentioned in terms of the correlations between variates and ranks in Equation 4.27. Specifically, the variant rank correlations for the uniform and logistic distributions are $\rho_{U_i,R(U_i)} = \sqrt{(n-1)/(n+1)}(1)$ and $\rho_{L_i,R(L_i)} = \sqrt{(n-1)/(n+1)}(3/\pi)$, respectively. And the variate rank correlation for a uniform-based polynomial is $\rho_{Y_i,R(Y_i)} = \rho_{Y_i,R(U_i)} = \sqrt{(n-1)/(n+1)}\sqrt{3/\pi}((1/35)(35c_{2i} + 63c_{4i} + 135c_{6i}))$. In terms of the logistic-based polynomial there is no convenient closed-form solution for the variate rank correlation $\rho_{Y_i,R(L_i)}$.

5

Other Transformations: The g-and-h and GLD Families of Distributions

5.1 Introduction

In this chapter we consider the g-and-h (e.g., Headrick, Kowalchuk, & Sheng 2008b; Kowalchuk & Headrick, 2009) and generalized lambda distribution (GLD; e.g., Ramberg, Dudewicz, Tadikamalla, & Mykytka, 1979; Karian & Dudewicz, 2000; Headrick & Mugdadi, 2006) families of transformations that are comparable to power method polynomials in terms of computational efficiency. As indicated in Chapter 1, these transformations also produce nonnormal distributions and have pdfs and cdfs of the general form in Equations 1.1 and 1.2. Moreover, like power method transformations, these alternatives can be used in the context of parameter estimation, distribution fitting, or for simulating univariate or multivariate nonnormal distributions.

We will begin with providing the requisite information for the univariate g-and-h and GLD families of transformations. Numerical examples are provided to demonstrate the two transformations in terms of computing cumulants, measures of central tendency, quantiles, distribution fitting, and in terms of graphing pdfs and cdfs. It is also demonstrated how the g-and-h, GLD, and power method transformations can be used to simulate or model combined data sets when only the mean, variance, skew, and kurtosis associated with the underlying individual data sets are available; that is, the raw data points are not available or accessible.

Extensions from univariate to multivariate data generation are presented with examples provided based on both g-and-h and GLD procedures. Finally, it is demonstrated that this methodology is general enough to allow for g-and-h, GLD, and power method transformations to be used together in the context of simulating correlated nonnormal distributions with specified cumulants.

5.2 The *g-and-h* Family

The *g-and-h* family of distributions is based on three transformations that produce nonnormal distributions with defined or undefined moments. Like the power method, these transformations are computationally efficient because they only require the knowledge of the *g* and *h* parameters and an algorithm that generates standard normal pseudorandom deviates (Z). Let the analytical and empirical forms of the quantile function for *g-and-h* distributions be defined as in Headrick et al. (2008b):

$$q_{g,h}(z) = g^{-1}(\exp\{gz\} - 1)\exp\{hz^2/2\} \tag{5.1}$$

$$q_{g,h}(Z) = g^{-1}(\exp\{gZ\} - 1)\exp\{hZ^2/2\} \tag{5.2}$$

where $q_{g,h}(z)$ is a strictly increasing monotonic function in z and where g and h are real numbers subject to the conditions that $g \neq 0$ and $h > 0$. The parameter $\pm g$ controls the skew of a distribution in terms of both direction and magnitude. The parameter h controls the tail-weight or elongation of a distribution and is positively related with kurtosis.

Two additional subclasses of distributions based on Equation 5.1 are the g and the h classes, which are defined as

$$q_{g,0}(z) = \lim_{h \to 0} q_{g,h}(z) = g^{-1}(\exp\{gz\} - 1) \tag{5.3}$$

$$q_{0,h}(z) = \lim_{g \to 0} q_{g,h}(z) = z\exp\{hz^2/2\} \tag{5.4}$$

where Equations 5.3 and 5.4 are lognormal (g) and symmetric (h) distributions, respectively. The explicit forms of the derivatives associated with Equations 5.1, 5.3, and 5.4 are

$$q'_{g,h}(z) = \exp\{gz + hz^2/2\} + g^{-1}(\exp\{hz^2/2\}(\exp\{gz\} - 1))hz \tag{5.5}$$

$$q'_{g,0}(z) = \lim_{h \to 0} q'_{g,h}(z) = \exp\{gz\} \tag{5.6}$$

$$q'_{0,h}(z) = \lim_{g \to 0} q'_{g,h}(z) = \exp\{hz^2/2\}(1 + hz^2) \tag{5.7}$$

Thus, using Equations 1.1 and 1.2, the parametric forms ($\mathbb{R}^2$) of the pdfs and cdfs for the *g-and-h*, *g*, and *h* distributions are

$$f_{q_{g,h}(Z)}(q_{g,h}(z)) = f_{q_{g,h}(Z)}(q_{g,h}(x,y)) = f_{q_{g,h}(Z)}\left(q_{g,h}(z), \frac{f_Z(z)}{q'_{g,h}(z)}\right) \tag{5.8}$$

$$F_{q_{g,h}(Z)}(q_{g,h}(z)) = F_{q_{g,h}(Z)}(q_{g,h}(x,y)) = F_{q_{g,h}(Z)}(q_{g,h}(z), F_Z(z)) \tag{5.9}$$

$$f_{q_{g,0}(Z)}(q_{g,0}(z)) = f_{q_{g,0}(Z)}(q_{g,0}(x,y)) = f_{q_{g,0}(Z)}\left(q_{g,0}(z), \frac{f_Z(z)}{q'_{g,0}(z)}\right) \tag{5.10}$$

$$F_{q_{g,0}(Z)}(q_{g,0}(z)) = F_{q_{g,0}(Z)}(q_{g,0}(x,y)) = F_{q_{g,0}(Z)}(q_{g,0}(z), F_Z(z)) \tag{5.11}$$

$$f_{q_{0,h}(Z)}(q_{0,h}(z)) = f_{q_{0,h}(Z)}(q_{0,h}(x,y)) = f_{q_{0,h}(Z)}\left(q_{0,h}(z), \frac{f_Z(z)}{q'_{0,h}(z)}\right) \tag{5.12}$$

$$F_{q_{0,h}(Z)}(q_{0,h}(z)) = F_{q_{0,h}(Z)}(q_{0,h}(x,y)) = F_{q_{0,h}(Z)}(q_{0,h}(z), F_Z(z)) \tag{5.13}$$

where $f_z(z)$ and $F_z(z)$ are the standard normal pdf and cdf.

For any *g*-and-*h* distribution, the *k*-th moment will exist if we meet the condition that $0 \le h < 1/k$ (e.g., Headrick et al., 2008b). As such, if the first four moments are defined (i.e., $0 \le h < 0.25$), then the mean (μ), variance (σ^2), skew (γ_3), and kurtosis (γ_4) for *g*-and-*h* distributions are (Kowalchuck & Headrick, 2009)

$$\mu(g,h) = (\exp\{g^2(2-2h)^{-1}\}-1)/(g(1-h)^{\frac{1}{2}}) \tag{5.14}$$

$$\sigma^2(g,h) = [(1-2\exp\{g^2(2-4h)^{-1}\}+\exp\{2g^2(1-2h)^{-1}\})/(1-2h)^{\frac{1}{2}} \\ + (\exp\{g^2(2-2h)^{-1}\}-1)^2/(h-1)]/g^2 \tag{5.15}$$

Equation 5.16 should appear as:

$$\gamma_3(g,h) = [(3\exp\{g^2/(2-6h)\}+\exp\{9g^2/(2-6h)\}-3\exp\{2g^2/(1-3h)\}-1) \\ \div (1-3h)^{\frac{1}{2}}-3(1-2\exp\{g^2/(2-4h)\}+\exp\{2g^2/(1-2h)\})\ (\exp\{g^2 \\ \div(2-2h)\}-1)/((1-2h)^{\frac{1}{2}}(1-h)^{\frac{1}{2}})+2(\exp\{g^2/(2-2h)\}-1)^3/\ (1-h)^{\frac{3}{2}}] \\ \div [g^3(((1-2\exp\{g^2/(2-4h)\}+\exp\{2g^2/(1-2h)\})/(1-2h)^{\frac{1}{2}} \\ + (\exp\{g^2/(2-2h)\}-1)^2/(h-1))/g^2)^{\frac{3}{2}}] \tag{5.16}$$

and Equation 5.17 should appear as:

$$\gamma_4(g,h) = [\exp\{8g^2/(1-4h)\}(1+6\exp\{6g^2/(4h-1)\}+\exp\{8g^2/(4h-1)\} \\ -4\exp\{7g^2/(8h-2)\}-4\exp\{15g^2/(8h-2)\})/(1-4h)^{\frac{1}{2}}-4(3\exp\{g^2 \\ \div(2-6h)\}+\exp\{9g^2/(2-6h)\}-3\exp\{2g^2/(1-3h)\}-1)(\exp\{g^2 \\ \div(2-2h)\}-1)/((1-3h)^{\frac{1}{2}}(1-h)^{\frac{1}{2}})-6(\exp\{g^2/(2-2h)\}-1)^4/(h-1)^2 \\ -12(1-2\exp\{g^2/(2-4h)\}+\exp\{2g^2/(1-2h)\})(\exp\{g^2/(2-2h)\}-1)^2 \\ \div((1-2h)^{\frac{1}{2}}(h-1))+3(1-2\exp\{g^2/(2-4h)\}\exp\{2g^2/(1-2h)\})^2 \\ \div(2h-1)]/[(1-2\exp\{g^2/(2-4h)\}+\exp\{2g^2/(1-2h)\})/(1-2h)^{\frac{1}{2}} \\ +(\exp\{g^2/(2-2h)\}-1)^2/(h-1)]^2. \tag{5.17}$$

Simultaneously solving Equations 5.16 and 5.17 for specified values of skew and kurtosis will give the solutions of the g and h parameters for Equations 5.1 and 5.2. The mean and variance for a distribution can be determined by evaluating Equations 5.14 and 5.15 using the solutions of g and h.

Using 5.14–5.17, we have for g distributions, that is, as $h \to 0$:

$$\mu(g) = (\exp\{g^2/2\} - 1)/g \tag{5.18}$$

$$\sigma^2(g) = (\exp\{g^2\}(\exp\{g^2\} - 1))/g^2 \tag{5.19}$$

$$\gamma_3(g) = g(2 + \exp\{g^2\})(\exp\{-g^2/2\})\sqrt{g^{-2}\exp\{g^2\}(\exp\{g^2\} - 1)} \tag{5.20}$$

$$\gamma_4(g) = 3\exp\{2g^2\} + 2\exp\{3g^2\} + \exp\{4g^2\} - 6 \tag{5.21}$$

Given these equations, a user can specify one equation—either skew (Equation 5.20) or kurtosis (Equation 5.21)—and solve for the g parameter. Analogously, using Equation 5.14–5.17, for h distributions we have as $g \to 0$:

$$\mu(h) = 0 \tag{5.22}$$

$$\sigma^2(h) = 1/(1 - 2h)^{\frac{3}{2}} \tag{5.23}$$

$$\gamma_3(h) = 0 \tag{5.24}$$

$$\gamma_4(h) = 3(1 - 2h)^3(1/(1 - 4h)^{\frac{5}{2}} + 1/(2h - 1)^3) \tag{5.25}$$

Other measures of the central tendency associated with the g-and-h family, such as the mode, can be determined by solving for the critical value of z ($\bar{z}$) that maximizes the height of a pdf, for example, $f_Z(z)/q'_{g,h}(z)$ in Equation 5.8, which is analogous to what was shown in Property 2.5 for the power method. The modes are located at $q_{g,h}(\bar{z})$, $q_{g,0}(\bar{z})$, and $q_{0,h}(\bar{z}) = 0$, where the height of any h distribution will have the same height of the standard normal distribution $(2\pi)^{-\frac{1}{2}}$. The median for any g-and-h, g, or h distribution will always be zero, that is, $\lim_{z \to 0} q_{g,h}(z) = \text{median} = 0$, and a 100α symmetric trimmed mean (TM) can be computed as given in Headrick et al. (2008b):

$$\text{TM} = (1 - 2\alpha)^{-1} \int_{F_Z^{-1}(\alpha)}^{F_Z^{-1}(1-\alpha)} q_{g,h}(z)dF_Z(z) \tag{5.26}$$

We subsequently provide the basic univariate details for the GLD.

5.3 The Generalized Lambda Distributions (GLDs)

The class of GLDs was first proposed by Ramberg and Schmeiser (1972, 1974) as a generalization of Tukey's (1960) lambda distribution. The analytical and stochastic forms of the quantile function considered herein are expressed as

$$q(u) = \lambda_1 + (u^{\lambda_3} - (1-u)^{\lambda_4})/\lambda_2 \tag{5.27}$$

$$q(U) = \lambda_1 + (U^{\lambda_3} - (1-U)^{\lambda_4})/\lambda_2 \tag{5.28}$$

where pseudorandom deviates (U) are generated from a uniform distribution on the interval [0, 1]. The values of λ_1 and λ_2 are location and scale parameters, while λ_3 and λ_4 are shape parameters that determine the skew and kurtosis of a GLD in Equation 5.27 or Equation 5.28. We note that if $\lambda_3 \neq \lambda_4$ ($\lambda_3 = \lambda_4$), then the distribution is asymmetric (symmetric).

The derivative associated with Equation 5.27 is expressed as

$$q'(u) = (\lambda_3 u^{\lambda_3 - 1} + \lambda_4 (1-u)^{\lambda_4 - 1})/\lambda_2 \tag{5.29}$$

As such, using Equation 1.1 and Equation 1.2, the parametric forms ($\mathbb{R}^2$) of the pdf and cdf associated with Equation 5.27 are

$$f_{q(U)}(q(u)) = f_{q(U)}(q(x,y)) = f_{q(U)}\left(q(u), \frac{f_U(u)}{q'(u)}\right) = f_{q(U)}\left(q(u), \frac{1}{q'(u)}\right) \tag{5.30}$$

$$F_{q(U)}(q(u)) = F_{q(U)}(q(x,y)) = F_{q(U)}(q(u), F_U(u)) = F_{q(U)}(q(u), u) \tag{5.31}$$

since $f_U(u) = 1$ and $f_U(u) = u$ are the pdf and cdf for the regular uniform distribution. As with the power method and the *g-and-h* transformations, inspection of (5.30) indicates that the derivative $q'(u)$ must be positive to ensure a valid GLD pdf, that is, the quantile function $q(u)$ must be a strictly increasing function in u for all $0 \leq u \leq 1$. In particular, and as indicated by Karian and Dudewicz (2000, p. 12), a valid pdf will be produced if λ_2, λ_3, and λ_4 all have the same sign; that is, they are either all positive or all negative. For more specific details on the parameter space and moments for valid pdfs, see Karian and Dudewicz (2000), Karian, Dudewicz, and McDonald (1996), Ramberg and Schmeiser (1974), and Ramberg et al. (1979).

The formulae for the mean, variance, skew, and kurtosis for standardized GLDs are (Ramberg et al., 1979)

$$\gamma_1 = 0 = \lambda_2^{-1}((1+\lambda_3)^{-1} - (1+\lambda_4)^{-1}) + \lambda_1 \tag{5.32}$$

$$\gamma_2 = 1 = \lambda_2^{-2}(-2\beta(1+\lambda_3, 1+\lambda_4) + (1+2\lambda_3)^{-1} + (1+2\lambda_4)^{-1} - ((1+\lambda_3)^{-1} - (1+\lambda_4)^{-1})^2) \tag{5.33}$$

$$\gamma_3 = \lambda_2^{-3}(3\beta(1+\lambda_3,1+2\lambda_4) - 3\beta(1+2\lambda_3,1+\lambda_4) + (1+3\lambda_3)^{-1} - (1+3\lambda_4)^{-1}$$
$$+2((1+\lambda_3)^{-1} - (1+\lambda_4)^{-1})^3 - 3((1+\lambda_3)^{-1} - (1+\lambda_4)^{-1})(-2\beta(1+\lambda_3,1+\lambda_4)$$
$$+(1+2\lambda_3)^{-1} + (1+2\lambda_4)^{-1})) \tag{5.34}$$

$$\gamma_4 = \lambda_2^{-4}(-4\beta(1+\lambda_3,1+3\lambda_4) + 6\beta(1+2\lambda_3,1+2\lambda_4) - 4\beta(1+3\lambda_3,1+\lambda_4)$$
$$+ (1+4\lambda_3)^{-1} + (1+4\lambda_4)^{-1} - 3((1+\lambda_3)^{-1} - (1+\lambda_4)^{-1})^4 + 6((1+\lambda_3)^{-1} - (1+\lambda_4)^{-1})^2$$
$$\times (-2\beta(1+\lambda_3,1+\lambda_4) + (1+2\lambda_3)^{-1} + (1+2\lambda_4)^{-1}) - 4((1+\lambda_3)^{-1} - (1+\lambda_4)^{-1})$$
$$\times (3\beta(1+\lambda_3,1+2\lambda_4) - 3\beta(1+2\lambda_3,1+\lambda_4) + (1+3\lambda_3)^{-1} - (1+3\lambda_4)^{-1})) - 3 \tag{5.35}$$

where the beta function has the form $\beta(r,s) = \int_0^1 x^{r-1}(1-x)^{s-1}\, dx$ for $r, s > 0$. Simultaneously solving Equations 5.32–5.35 for specified values of skew and kurtosis provides the $\lambda_{1,2,3,4}$ parameters for Equations 5.27 and 5.28. We note that γ_4 (Equation 5.35) is scaled here so that, for example, kurtosis for the normal distribution is zero and not 3.

As with power method and g-and-h distributions, a mode (if it exists) can be determined by finding the critical value of u ($\bar{u}$) that maximizes the height of the pdf $1/q'(u)$ in Equation 5.30. The mode is located at $q(\bar{u})$. The median is determined by evaluating $\lim_{u \to 1/2} q(u) = \text{median} = (\lambda_1\lambda_2 + 2^{-\lambda_3} - 2^{-\lambda_4})/\lambda_2$, and a 100α symmetric trimmed mean is computed as

$$\text{TM} = (1-2\alpha)^{-1} \int_\alpha^{1-\alpha} q(u)du \tag{5.36}$$

In summary, the essential methodology was presented for the g-and-h and GLD families of distributions in terms of determining the (1) parameters (g, h, λ_i) for distributions when the first four moments (cumulants) are defined, (2) conditions for valid pdfs considered herein, and (3) measures of central tendency. In the next section, examples are provided to demonstrate the methodology in the context of graphing and fitting g-and-h (GLD) pdfs to data and computing various indices associated with these pdfs.

5.4 Numerical Examples

Presented in Tables 5.1–5.6 are some examples of pdfs and cdfs for the g-and-h and GLD families of distributions. Included in these tables are computations of the parameters (g, h, λ_i), quartiles, heights, and modes of the distributions.

TABLE 5.1

Examples of g-and-h PDFs and CDFs

Parameters	PDF	CDF

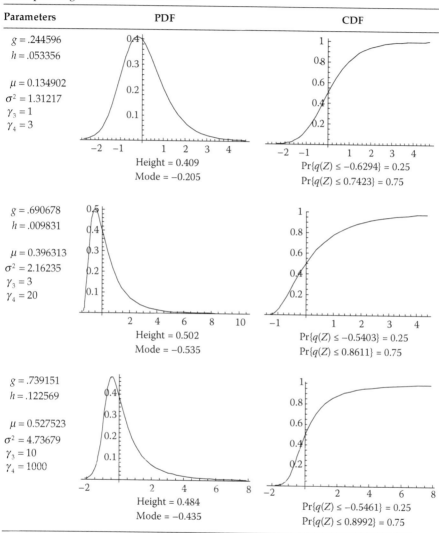

$g = .244596$
$h = .053356$

$\mu = 0.134902$
$\sigma^2 = 1.31217$
$\gamma_3 = 1$
$\gamma_4 = 3$

Height = 0.409
Mode = −0.205

Pr{q(Z) ≤ −0.6294} = 0.25
Pr{q(Z) ≤ 0.7423} = 0.75

$g = .690678$
$h = .009831$

$\mu = 0.396313$
$\sigma^2 = 2.16235$
$\gamma_3 = 3$
$\gamma_4 = 20$

Height = 0.502
Mode = −0.535

Pr{q(Z) ≤ −0.5403} = 0.25
Pr{q(Z) ≤ 0.8611} = 0.75

$g = .739151$
$h = .122569$

$\mu = 0.527523$
$\sigma^2 = 4.73679$
$\gamma_3 = 10$
$\gamma_4 = 1000$

Height = 0.484
Mode = −0.435

Pr{q(Z) ≤ −0.5461} = 0.25
Pr{q(Z) ≤ 0.8992} = 0.75

The computations and graphics were obtained using the *Mathematica* functions that were discussed in Section 2.4 for the power method.

Provided in Tables 5.7 and 5.8 are examples of g-and-h and GLD pdfs superimposed on histograms of circumference measures (centimeters) taken from the neck, chest, hip, and ankle of 252 adult men (http://lib.stat.cmu.edu/datasets/bodyfat). Inspection of these tables indicates that both the g-and-h and GLD pdfs provide good approximations to the empirical data.

TABLE 5.2

Examples of (Lognormal) g PDFs and CDFs

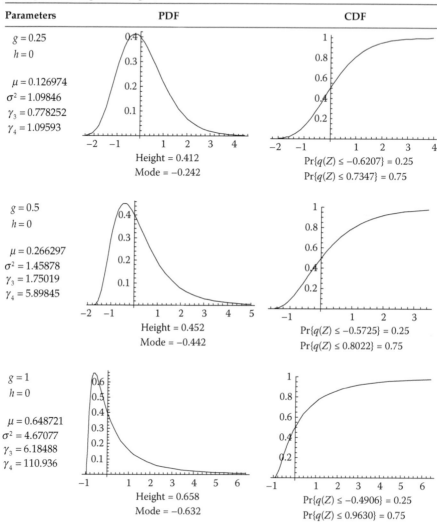

Parameters	PDF	CDF
$g = 0.25$ $h = 0$ $\mu = 0.126974$ $\sigma^2 = 1.09846$ $\gamma_3 = 0.778252$ $\gamma_4 = 1.09593$	Height = 0.412 Mode = -0.242	Pr$\{q(Z) \leq -0.6207\} = 0.25$ Pr$\{q(Z) \leq 0.7347\} = 0.75$
$g = 0.5$ $h = 0$ $\mu = 0.266297$ $\sigma^2 = 1.45878$ $\gamma_3 = 1.75019$ $\gamma_4 = 5.89845$	Height = 0.452 Mode = -0.442	Pr$\{q(Z) \leq -0.5725\} = 0.25$ Pr$\{q(Z) \leq 0.8022\} = 0.75$
$g = 1$ $h = 0$ $\mu = 0.648721$ $\sigma^2 = 4.67077$ $\gamma_3 = 6.18488$ $\gamma_4 = 110.936$	Height = 0.658 Mode = -0.632	Pr$\{q(Z) \leq -0.4906\} = 0.25$ Pr$\{q(Z) \leq 0.9630\} = 0.75$

We note that to fit the g-and-h distributions to the data, the following linear transformations had to be imposed on $q_{g,h}(z)$: $Aq_{g,h}(z) + B$ where $A = s/\sigma$ and $B = m - A\mu$, and where the values of the means (m, μ) and standard deviations (s, σ) for the data and g-and-h pdfs are given in Table 5.7.

Tables 5.9–5.11 give the chi-square goodness of fit statistics for the g-and-h, GLD, and (fifth-order) power method pdfs associated with the chest data from panel B of Tables 5.7 and 5.8. The asymptotic p-values of each procedure indicate good fits to the data. We note that the p-value is larger for the power

TABLE 5.3

Examples of Symmetric *h* PDFs and CDFs

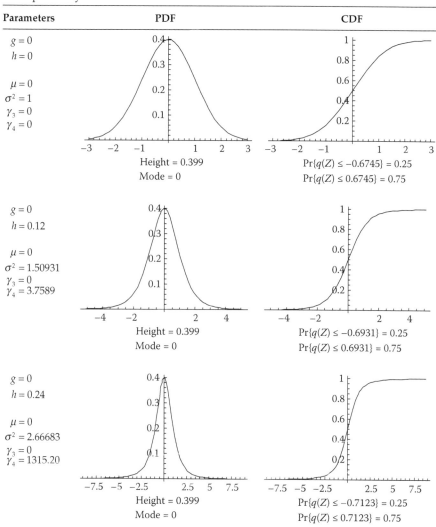

Parameters	PDF	CDF
$g = 0$ $h = 0$ $\mu = 0$ $\sigma^2 = 1$ $\gamma_3 = 0$ $\gamma_4 = 0$	Height = 0.399 Mode = 0	$\Pr\{q(Z) \le -0.6745\} = 0.25$ $\Pr\{q(Z) \le 0.6745\} = 0.75$
$g = 0$ $h = 0.12$ $\mu = 0$ $\sigma^2 = 1.50931$ $\gamma_3 = 0$ $\gamma_4 = 3.7589$	Height = 0.399 Mode = 0	$\Pr\{q(Z) \le -0.6931\} = 0.25$ $\Pr\{q(Z) \le 0.6931\} = 0.75$
$g = 0$ $h = 0.24$ $\mu = 0$ $\sigma^2 = 2.66683$ $\gamma_3 = 0$ $\gamma_4 = 1315.20$	Height = 0.399 Mode = 0	$\Pr\{q(Z) \le -0.7123\} = 0.25$ $\Pr\{q(Z) \le 0.7123\} = 0.75$

method than the *g*-and-*h* fit because two degrees of freedom are lost because two additional cumulants (fifth and sixth) are being matched. Further, the *g*-and-*h* and GLD trimmed means given in Table 5.12 also indicate a good fit, as the TMs are all within the 95% bootstrap confidence intervals based on the data. The confidence intervals are based on 25,000 bootstrap samples.

Another use of the *g*-and-*h*, GLD, or power method transformations is in the context of simulating combined data sets when the individual raw data sets are not accessible. Data that are considered not accessible may occur

TABLE 5.4

Examples of Asymmetric GLD PDFs and CDFs

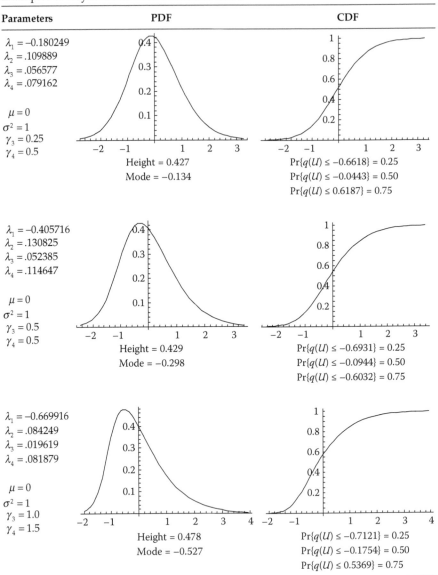

Parameters	PDF	CDF
$\lambda_1 = -0.180249$ $\lambda_2 = .109889$ $\lambda_3 = .056577$ $\lambda_4 = .079162$ $\mu = 0$ $\sigma^2 = 1$ $\gamma_3 = 0.25$ $\gamma_4 = 0.5$	Height = 0.427 Mode = −0.134	$\Pr\{q(U) \le -0.6618\} = 0.25$ $\Pr\{q(U) \le -0.0443\} = 0.50$ $\Pr\{q(U) \le 0.6187\} = 0.75$
$\lambda_1 = -0.405716$ $\lambda_2 = .130825$ $\lambda_3 = .052385$ $\lambda_4 = .114647$ $\mu = 0$ $\sigma^2 = 1$ $\gamma_3 = 0.5$ $\gamma_4 = 0.5$	Height = 0.429 Mode = −0.298	$\Pr\{q(U) \le -0.6931\} = 0.25$ $\Pr\{q(U) \le -0.0944\} = 0.50$ $\Pr\{q(U) \le -0.6032\} = 0.75$
$\lambda_1 = -0.669916$ $\lambda_2 = .084249$ $\lambda_3 = .019619$ $\lambda_4 = .081879$ $\mu = 0$ $\sigma^2 = 1$ $\gamma_3 = 1.0$ $\gamma_4 = 1.5$	Height = 0.478 Mode = −0.527	$\Pr\{q(U) \le -0.7121\} = 0.25$ $\Pr\{q(U) \le -0.1754\} = 0.50$ $\Pr\{q(U) \le 0.5369\} = 0.75$

either for legal reasons, such as with medical or educational data, or just simply because the data are lost or have been destroyed. For example, suppose there are several (*k*) school districts within a certain region (e.g., a province or state) where each district has a particular test score for each student. However, for legal reasons, the individual test scores themselves cannot be

TABLE 5.5

Examples of Asymmetric GLD PDFs and CDFs

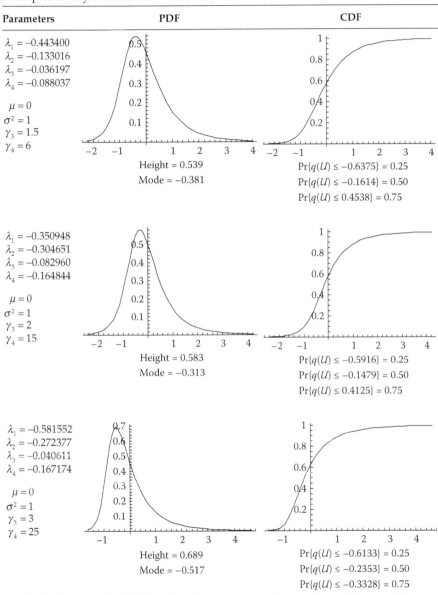

Parameters	PDF	CDF
$\lambda_1 = -0.443400$ $\lambda_2 = -0.133016$ $\lambda_3 = -0.036197$ $\lambda_4 = -0.088037$ $\mu = 0$ $\sigma^2 = 1$ $\gamma_3 = 1.5$ $\gamma_4 = 6$	Height = 0.539 Mode = −0.381	$\Pr\{q(U) \leq -0.6375\} = 0.25$ $\Pr\{q(U) \leq -0.1614\} = 0.50$ $\Pr\{q(U) \leq 0.4538\} = 0.75$
$\lambda_1 = -0.350948$ $\lambda_2 = -0.304651$ $\lambda_3 = -0.082960$ $\lambda_4 = -0.164844$ $\mu = 0$ $\sigma^2 = 1$ $\gamma_3 = 2$ $\gamma_4 = 15$	Height = 0.583 Mode = −0.313	$\Pr\{q(U) \leq -0.5916\} = 0.25$ $\Pr\{q(U) \leq -0.1479\} = 0.50$ $\Pr\{q(U) \leq 0.4125\} = 0.75$
$\lambda_1 = -0.581552$ $\lambda_2 = -0.272377$ $\lambda_3 = -0.040611$ $\lambda_4 = -0.167174$ $\mu = 0$ $\sigma^2 = 1$ $\gamma_3 = 3$ $\gamma_4 = 25$	Height = 0.689 Mode = −0.517	$\Pr\{q(U) \leq -0.6133\} = 0.25$ $\Pr\{q(U) \leq -0.2353\} = 0.50$ $\Pr\{q(U) \leq -0.3328\} = 0.75$

made public, and the only indices available are descriptive statistics, such as the mean, median, standard deviation, skew, and kurtosis. Further, suppose our objective is to obtain an approximation of the distribution of test scores for the entire region, but we only have the descriptive statistics from

TABLE 5.6

Examples of Symmetric GLD PDFs and CDFs

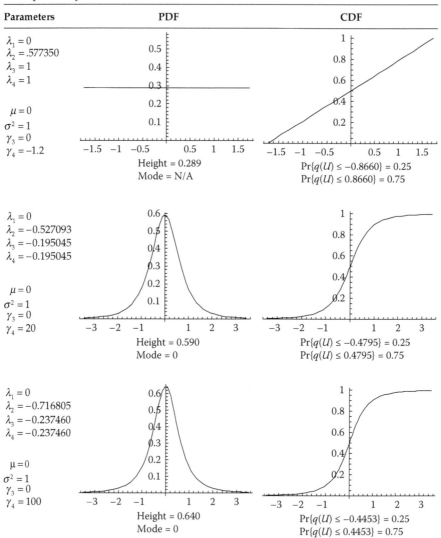

Parameters	PDF	CDF
$\lambda_1 = 0$ $\lambda_2 = .577350$ $\lambda_3 = 1$ $\lambda_4 = 1$ $\mu = 0$ $\sigma^2 = 1$ $\gamma_3 = 0$ $\gamma_4 = -1.2$	Height = 0.289 Mode = N/A	Pr$\{q(U) \le -0.8660\} = 0.25$ Pr$\{q(U) \le 0.8660\} = 0.75$
$\lambda_1 = 0$ $\lambda_2 = -0.527093$ $\lambda_3 = -0.195045$ $\lambda_4 = -0.195045$ $\mu = 0$ $\sigma^2 = 1$ $\gamma_3 = 0$ $\gamma_4 = 20$	Height = 0.590 Mode = 0	Pr$\{q(U) \le -0.4795\} = 0.25$ Pr$\{q(U) \le 0.4795\} = 0.75$
$\lambda_1 = 0$ $\lambda_2 = -0.716805$ $\lambda_3 = -0.237460$ $\lambda_4 = -0.237460$ $\mu = 0$ $\sigma^2 = 1$ $\gamma_3 = 0$ $\gamma_4 = 100$	Height = 0.640 Mode = 0	Pr$\{q(U) \le -0.4453\} = 0.25$ Pr$\{q(U) \le 0.4453\} = 0.75$

each district within the region. To accomplish our objective, let us define the first two districts within the region as having test scores denoted as $X_1 = \{x_{11}, x_{12}, \ldots, x_{1m}\}$ and $X_2 = \{x_{21}, x_{22}, \ldots, x_{2n}\}$. The two districts have m and n test scores where we only know the means, variances, skew, and kurtosis of these two sets of scores, which are denoted as M_i, V_i, S_i, and K_i for $i = 1,2$. Let A_1 denote the combined data of X_1 and X_2 as $A_1 = X_1 \cup X_2$ or $A_1 = \{x_{11}, x_{12}, \ldots, x_{1m}, x_{21}, x_{22}, \ldots, x_{2n}\}$. The mean, variance, skew, and kurtosis associated with

TABLE 5.7

Examples of *g*-and-*h* PDF Approximations to Empirical PDFs Using Measures of Circumference (Centimeters) Taken from 252 Men

Data	Parameters	
$m = 37.992$ $s = 2.431$ $\gamma_3 = 0.549$ $\gamma_4 = 2.642$	$\mu = 0.065$ $\sigma = 1.172$ $g = 0.113318$ $h = 0.088872$	
$m = 100.824$ $s = 8.430$ $\gamma_3 = 0.677$ $\gamma_4 = 0.944$	$\mu = 0.108$ $\sigma = 1.052$ $g = 0.209937$ $h = 0.010783$	
$m = 99.905$ $s = 7.164$ $\gamma_3 = 1.488$ $\gamma_4 = 7.300$	$\mu = 0.172$ $\sigma = 1.248$ $g = 0.293304$ $h = 0.085829$	
$m = 23.102$ $s = 1.695$ $\gamma_3 = 2.242$ $\gamma_4 = 11.686$	$\mu = 0.292$ $\sigma = 1.321$ $g = 0.512894$ $h = 0.038701$	

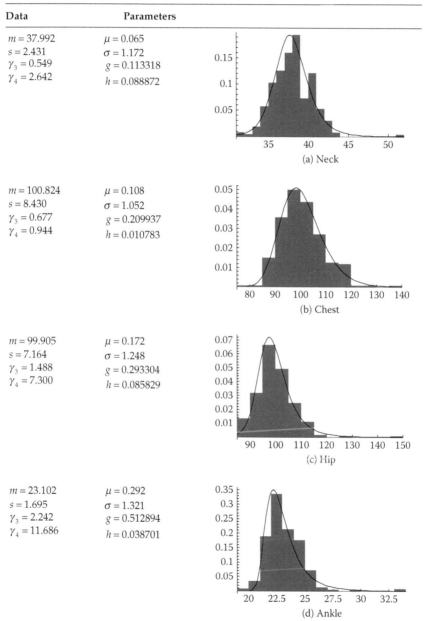

(a) Neck

(b) Chest

(c) Hip

(d) Ankle

Note: The *g*-and-*h* pdfs were scaled using $Aq(z) + B$, where $A = s/\sigma$ and $B = m - A\mu$.

TABLE 5.8

Examples of GLD Approximations to Empirical PDFs Using Measures of Circumference (Centimeters) Taken from 252 Men

Data	Parameters	
$m = 37.992$ $s = 2.431$ $\gamma_3 = 0.549$ $\gamma_4 = 2.642$	$\lambda_1 = -0.174953$ $\lambda_3 = -0.044428$	$\lambda_2 = -0.105919$ $\lambda_4 = -0.061055$

(a) Neck

Data	Parameters	
$m = 100.824$ $s = 8.430$ $\gamma_3 = 0.677$ $\gamma_4 = 0.944$	$\lambda_1 = -0.459489$ $\lambda_3 = 0.032539$	$\lambda_2 = 0.092535$ $\lambda_4 = 0.079952$

(b) Chest

Data	Parameters	
$m = 99.905$ $s = 7.164$ $\gamma_3 = 1.488$ $\gamma_4 = 7.300$	$\lambda_1 = -0.361384$ $\lambda_3 = -0.058017$	$\lambda_2 = -0.195422$ $\lambda_4 = -0.116774$

(c) Hip

Data	Parameters	
$m = 23.102$ $s = 1.695$ $\gamma_3 = 2.242$ $\gamma_4 = 11.686$	$\lambda_1 = -0.587304$ $\lambda_3 = -0.030796$	$\lambda_2 = -0.174614$ $\lambda_4 = -0.118419$

(d) Ankle

TABLE 5.9

Observed and Expected Frequencies and Chi-Square Test Based on the *g*-and-*h*
Approximation to the Chest Data in Figure 5.1

Cumulative %	*g*-and-*h* Class Intervals	Observed Freq	Expected Freq
5	<88.70	12	12.60
10	88.70–90.89	13	12.60
15	90.89–92.47	13	12.60
30	92.47–95.98	35	37.80
50	95.98–99.96	56	50.40
70	99.96–104.40	49	50.40
85	104.40–109.28	39	37.80
90	109.28–111.83	9	12.60
95	111.83–115.90	13	12.60
100	>115.90	13	12.60
$\chi^2 = 2.015$	$\Pr\{\chi_5^2 \le 2.015\} = .153$	$n = 252$	

A_1 can be determined as

$$M_{A_1} = \frac{m}{m+n} M_1 + \frac{n}{m+n} M_2 \tag{5.37}$$

$$V_{A_1} = \frac{m^2 V_1 + n^2 V_2 - nV_1 - nV_2 - mV_1 - mV_2 + mnV_1 + mnV_2 + mn(M_1 - M_2)^2}{(n+m-1)(n+m)} \tag{5.38}$$

TABLE 5.10

Observed and Expected Frequencies and Chi-Square Test Based on the GLD
Approximation to the Chest Data in Figure 5.1

Cumulative %	GLD Class Intervals	Observed Freq	Expected Freq
5	<88.86	13	12.60
10	88.86–91.14	13	12.60
15	91.14–92.67	13	12.60
30	92.67–96.01	35	37.80
50	96.01–99.83	55	50.40
70	99.83–104.26	47	50.40
85	104.26–109.29	41	37.80
90	109.29–111.96	9	12.60
95	111.96–116.20	13	12.60
100	>116.20	13	12.60
$\chi^2 = 2.220$	$\Pr\{\chi_5^2 \le 2.015\} = .182$	$n = 252$	

TABLE 5.11

Observed and Expected Frequencies and Chi-Square Test Based on the Power Method Approximation to the Chest Data in Figure 5.1

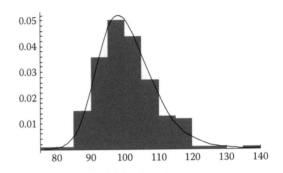

Cumulative %	g-and-h Class Intervals	Observed Freq	Expected Freq
5	<88.79	13	12.60
10	88.79–91.01	12	12.60
15	91.01–92.55	13	12.60
30	92.55–95.98	35	37.80
50	95.98–99.90	56	50.40
70	99.90–104.34	49	50.40
85	104.34–109.29	39	37.80
90	109.29–111.90	9	12.60
95	111.90–116.05	13	12.60
100	>116.05	13	12.60
$\chi^2 = 2.015$	$\Pr\{\chi_3^2 \le 2.015\} = .431$	$n = 252$	

$$S_{A_1} = (m+n)\left[\frac{(m-2)(m-1)S_1 V_1^{\frac{3}{2}}}{m} + \frac{(n-2)(n-1)S_2 V_2^{\frac{3}{2}}}{n} + 3(m-1)V_1 M_1\right.$$

$$+\ mM_1^3 + 3(n-1)V_2 M_2 + nM_2^3 + \frac{2(mM_1 + nM_2)^3}{(m+n)^2}$$

$$\left. -\ \frac{3(mM_1 + nM_2)((m-1)V_1 + (n-1)V_2 + mM_1^2 + nM_2^2)}{m+n}\right]$$

$$\left((m+n-2)(m+n-1)V_{A_1}^{\frac{3}{2}}\right) \tag{5.39}$$

TABLE 5.12

Examples of g-and-*h* and GLD Trimmed Means (TM)

Empirical Distribution	20% TM	g-and-*h* TM	GLD TM
Neck	37.929 (37.756, 38.100)	37.899	37.872
Chest	100.13 (99.54, 100.75)	99.825	100.067
Hip	99.328 (98.908, 99.780)	99.020	99.099
Ankle	22.914 (22.798, 23.007)	22.800	22.802

Note: Each TM is based on a sample size of $n = 152$ and has a 95% bootstrap confidence interval enclosed in parentheses.

$$K_{A_1} = \frac{1}{(m+n-3)(m+n-2)}\left\{(m+n-1)\left[-3(m+n-1)+((m+n)^3(m+n+1)\right.\right.$$

$$\times\left(\frac{1}{m(m+1)}\left((K_1(m-3)(m-2)+3(m-1)^2)(m-1)V_1^2\right)\right.$$

$$+\frac{1}{n(n+1)}\left((K_2(n-3)(n-2)+3(n-1)^2)(n-1)V_2^2\right)$$

$$+\frac{4(m-2)(m-1)S_1V_1^{\frac{3}{2}}M_1}{m}+6(m-1)V_1M_1^2+mM_1^4$$

$$+\frac{4(n-2)(n-1)S_2V_2^{\frac{3}{2}}M_2}{n}+6(n-1)V_2M_2^2+nM_2^4-\frac{3(mM_1+nM_2)^4}{(m+n)^3}$$

$$+\frac{1}{(m+n)^2}\left(6(mM_1+nM_2)^2((m-1)V_1+(n-1)V_2+mM_1^2+nM_2^2)\right)$$

$$-\frac{1}{m+n}\left(4(mM_1+nM_2)\left(\frac{(m-2)(m-1)S_1V_1^{\frac{3}{2}}}{m}+\frac{(n-2)(n-1)S_2V_2^{\frac{3}{2}}}{n}\right.\right.$$

$$\left.\left.\left.+3(m-1)V_1M_1+mM_1^3+3(n-1)V_2M_2+nM_2^3\right)\right)\right)$$

$$\left.\left.\left((m+n-1)(m+n))^2V_{A_1}^2\right)\right]\right\}$$

(5.40)

The formulae associated with Equations 5.37–5.40 are general to the extent that if we wanted to combine all k districts in the region, then all we need to do is apply the formulae $k - 1$ times as $A_1 = X_1 \cup X_2$, $A_2 = A_1 \cup X_3$, $A_3 = A_2 \cup X_4$, ..., $A_{k-1} = A_{k-2} \cup X_k$. We note that Equations 5.37–5.40 will yield the exact same values of the statistics that any commonly used software package such as SPSS, SAS, Minitab, and so on, would yield if one actually had the combined raw data points.

TABLE 5.13

Sample Statistics from Two Data Sets (X_1, X_2) and the Combined Data (A_1)

Descriptives	X_1	X_2	$A_1 = X_1 \cup X_2$
Sample size	252	248	500
Mean	32.2734	28.6617	30.48200
Variance	9.12810	4.13120	9.903812
Skew	0.285530	−0.215601	0.530066
Kurtosis	0.498498	0.835870	0.602903

To demonstrate, suppose we have two data sets (X_1 and X_2) with the descriptive statistics listed in Table 5.13. Applying Equations 5.37–5.40 we get the statistics for the combined data listed under $A_1 = X_1 \cup X_2$. Using these statistics, presented in Table 5.14 are *g*-and-*h*, GLD, and power method approximations to the actual combined data. Inspection of these pdfs and cdfs indicates that the three transformations will produce similar approximations for this particular set of sample statistics, given in Table 5.13.

5.5 Multivariate Data Generation

The method for simulating correlated nonnormal distributions from either the *g*-and-*h* or GLD families follows the logic and steps described in Section 2.6 and the *Mathematica* source code given in Table 2.5 for power method polynomials. The basic difference is that the transformations from the *g*-and-*h* or GLD families will appear in place of the polynomials in Table 2.5. Further, if we are considering GLDs, then the standard normal cdf would also have to be included and used analogously, as in the source code of Table 2.5 for the logistic-based polynomial.

More specifically, suppose we desire to generate the *g*-and-*h* and GLD distributions given in Table 5.15 with the specified correlation matrix given in Table 5.16.

Using the methodology from the previous section, the solutions for the *g*-and-*h* (GLD) parameters are listed in Table 5.17. The intermediate correlation matrices associated with the specified correlation matrix are given in Tables 5.18 and 5.19 for the *g*-and-*h* and GLD transformations, respectively.

An example of solving an intermediate correlation for two *g*-and-*h* distributions is provided in Table 5.20. Note that the *g*-and-*h* distributions are standardized prior to solving for the correlation. Similarly, source code is also provided for the user in Table 5.21 for computing an intermediate correlation for two GLDs.

The other intermediate correlations are determined by applying the source code (independently) to the appropriate *g*-and-*h* or GLD parameters listed in

TABLE 5.14

The *g*-and-*h*, GLD, and Power Method Approximations of the Combined Data

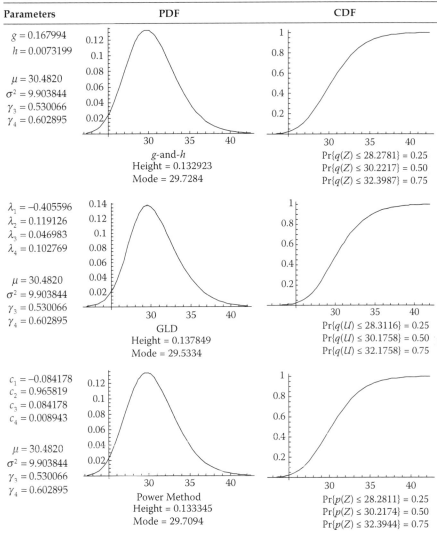

Parameters	PDF	CDF
$g = 0.167994$ $h = 0.0073199$ $\mu = 30.4820$ $\sigma^2 = 9.903844$ $\gamma_3 = 0.530066$ $\gamma_4 = 0.602895$	*g*-and-*h* Height = 0.132923 Mode = 29.7284	$\Pr\{q(Z) \leq 28.2781\} = 0.25$ $\Pr\{q(Z) \leq 30.2217\} = 0.50$ $\Pr\{q(Z) \leq 32.3987\} = 0.75$
$\lambda_1 = -0.405596$ $\lambda_2 = 0.119126$ $\lambda_3 = 0.046983$ $\lambda_4 = 0.102769$ $\mu = 30.4820$ $\sigma^2 = 9.903844$ $\gamma_3 = 0.530066$ $\gamma_4 = 0.602895$	GLD Height = 0.137849 Mode = 29.5334	$\Pr\{q(U) \leq 28.3116\} = 0.25$ $\Pr\{q(U) \leq 30.1758\} = 0.50$ $\Pr\{q(U) \leq 32.1758\} = 0.75$
$c_1 = -0.084178$ $c_2 = 0.965819$ $c_3 = 0.084178$ $c_4 = 0.008943$ $\mu = 30.4820$ $\sigma^2 = 9.903844$ $\gamma_3 = 0.530066$ $\gamma_4 = 0.602895$	Power Method Height = 0.133345 Mode = 29.7094	$\Pr\{p(Z) \leq 28.2811\} = 0.25$ $\Pr\{p(Z) \leq 30.2174\} = 0.50$ $\Pr\{p(Z) \leq 32.3944\} = 0.75$

Table 5.17 for the other pairwise combinations of correlations. The results of the Cholesky decompositions on the intermediate correlation matrices are listed in Tables 5.22 and 5.23 for the *g*-and-*h* distributions and the GLDs, respectively.

Using the results in Tables 5.22 and 5.23, equations of the form in Equation 2.56 are used to generate standard normal deviates (Z_i) correlated at the intermediate levels listed in Tables 5.18 and 5.19. In terms of the *g*-and-*h* transformation, the values of Z_i are then used in equations of the form in Equation 5.2 to generate the specified distributions in Table 5.15 with the

TABLE 5.15

Specified g-and-h ($q_{g,h}(Z_i)$) and GLD ($q(U_i)$) Distributions

Distribution	μ	σ^2	γ_3	γ_4
$q(Z_1)$	0.396313	2.162347	3	20
$q(Z_2)$	0.149607	1.670211	1.5	10
$q(Z_3)$	0.077467	1.533135	0.75	5
$q(Z_4)$	0.034843	1.188599	0.25	1
$q(U_1)$	0	1	0	0
$q(U_2)$	0	1	0.5	1
$q(U_3)$	0	1	1	3
$q(U_4)$	0	1	2	9

TABLE 5.16

Specified Correlation Matrix for the
g-and-h (GLD) Distributions in Table 5.15

	1	2	3	4
1	1			
2	0.30	1		
3	0.40	0.60	1	
4	0.50	0.70	0.80	1

TABLE 5.17

Solutions for g-and-h and GLD (λ_i) Parameters for the
Distributions in Table 5.15

Distribution	g	h
$q(Z_1)$	0.690678	0.009831
$q(Z_2)$	0.244996	0.114804
$q(Z_3)$	0.128290	0.115470
$q(Z_4)$	0.064220	0.052304

	λ_1	λ_2	λ_3	λ_4
$q(U_1)$	0	0.197451	0.134912	0.134912
$q(U_2)$	-0.291041	0.0603855	0.0258673	0.0447025
$q(U_3)$	-0.378520	-0.0561934	-0.0187381	-0.0388000
$q(U_4)$	-0.579659	-0.142276	-0.0272811	-0.0995192

TABLE 5.18

Intermediate Correlation Matrix for the Specified
g-and-h Correlations in Table 5.16

	Z_1	Z_2	Z_3	Z_4
Z_1	1			
Z_2	0.341860	1		
Z_3	0.450578	0.620528	1	
Z_4	0.561123	0.720648	0.809585	1

TABLE 5.19

Intermediate Correlation Matrix for the Specified
GLD Correlations in Table 5.16

	Z_1	Z_2	Z_3	Z_4
Z_1	1			
Z_2	0.302245	1		
Z_3	0.409561	0.610044	1	
Z_4	0.536005	0.732598	0.823539	1

TABLE 5.20

Mathematica Code for Solving an Intermediate Correlation in Table 5.18

(* The specified correlation between the g-and-h distributions $q_1(Z_1)$ and $q_2(Z_2)$
is 0.30. The solved intermediate correlation is $\rho_{z_1 z_2} = 0.34186$—see below.*)

$g_1 = .690678$
$h_1 = .009831$
$g_2 = .244996$
$h_2 = .114804$
$\mu_1 = 0.396313$
$\mu_2 = 0.149607$
$\sigma_1 = 1.470492$
$\sigma_2 = 1.292366$

$\rho_{z_1 z_2} = 0.341860$

$q_1 = g_1^{-1}(\exp\{g_1 z_1\} - 1)\exp\{h_1 z_1^2 / 2\}$
$q_2 = g_2^{-1}(\exp\{g_2 z_2\} - 1)\exp\{h_2 z_2^2 / 2\}$
$x_1 = (q_1 - \mu_1)/\sigma_1$
$x_2 = (q_2 - \mu_2)/\sigma_2$

$f_{12} = (2\pi\sqrt{1 - \rho_{z_1 z_2}^2})^{-1} \exp\{(-2(1 - \rho_{z_1 z_2}^2)^{-1}) z_1^2 (-2\rho_{z_1 z_2} z_1 z_2 + z_2^2)\}$

int = NIntegrate $[(x_1 x_2) f_{12}, \{z_1 - 8, 8\}, \{z_2; -8, 8\}, \text{Method} \to \text{Trapezoidal}]$

Solution: int = 0.300000

TABLE 5.21

Mathematica Code for Solving for an Intermediate Correlation in
Table 5.19

(* The specified correlation between the GLDs $q_1(U_1)$ and $q_2(U_2)$ is 0.30.
The solved intermediate correlation is $\rho_{z_1 z_2} = 0.302245$—see below.*)

$\lambda_{11} = 0.0$

$\lambda_{12} = 0.197451$

$\lambda_{13} = 0.134912$

$\lambda_{14} = 0.134912$

$\lambda_{21} = -0.291041$

$\lambda_{22} = 0.0603855$

$\lambda_{23} = 0.0258673$

$\lambda_{24} = 0.0447025$

$\rho_{z_1 z_2} = 0.302245$

$$\phi_1 = \int_{-\infty}^{z_1} (\sqrt{2\pi})^{-1} e^{-u_1^2/2} \, du_1$$

$$\phi_2 = \int_{-\infty}^{z_2} (\sqrt{2\pi})^{-1} e^{-u_2^2/2} \, du_2$$

$$x_1 = \lambda_{11} + (\phi_1^{\lambda_{13}} - (1-\phi_1)^{\lambda_{14}})/\lambda_{12}$$

$$x_2 = \lambda_{21} + (\phi_2^{\lambda_{23}} - (1-\phi_2)^{\lambda_{24}})/\lambda_{22}$$

$$f_{12} = (2\pi\sqrt{1-\rho_{z_1 z_2}^2})^{-1} \exp(-(2(1-\rho_{z_1 z_2}^2))^{-1}(z_1^2 - 2\rho_{z_1 z_2} z_1 z_2 + z_2^2))$$

int = NIntegrate $[(x_1 x_2) f_{12}, \{z_1 \ -8, 8\}, \{z_2, \ -8, 8\}$, Method $\rightarrow$ Trapezoidal]

Solution: int = 0.300000

TABLE 5.22

Cholesky Decomposition on the *g*-and-*h* Intermediate
Correlations in Table 5.18

$a_{11} = 1$	$a_{12} = 0.341860$	$a_{13} = 0.450578$	$a_{14} = 0.561123$
	$a_{22} = 0.939751$	$a_{23} = 0.496401$	$a_{24} = 0.562726$
		$a_{33} = 0.742001$	$a_{34} = 0.373877$
			$a_{44} = 0.478222$

TABLE 5.23

Cholesky Decomposition on the Intermediate GLD
Correlations in Table 5.19

$a_{11} = 1$	$a_{12} = 0.302245$	$a_{13} = 0.409561$	$a_{14} = 0.536005$
	$a_{22} = 0.953230$	$a_{23} = 0.510114$	$a_{24} = 0.598589$
		$a_{33} = 0.756335$	$a_{34} = 0.394882$
			$a_{44} = 0.445486$

TABLE 5.24

Empirical Estimates of the Specified Parameters in Table 5.15 Using Single Draws
of Size $N = 1,000,000$

Distribution	μ	σ^2	γ_3	γ_4
$q(Z_1)$	0.399968	2.18174	3.04910	20.9220
$q(Z_2)$	0.152857	1.67953	1.54206	10.3878
$q(Z_3)$	0.080527	1.53835	0.779421	5.17457
$q(Z_4)$	0.037412	1.19069	0.257674	1.01518
$q(U_1)$	−0.000267	1.00027	0.000381	0.996471
$q(U_2)$	0.000061	1.00015	0.503183	1.00451
$q(U_3)$	0.000273	0.999419	0.999521	3.00312
$q(U_4)$	−0.001382	1.00021	2.00419	9.00446

specified correlations in Table 5.16. In the context of the GLD, the Z_i are first
transformed to regular uniform deviates using Equation 2.57, which is subsequently used in equations of the form of Equation 5.28 to produce the GLDs
with specified cumulants and correlations given in Tables 5.15 and 5.16.

To empirically demonstrate, single draws of size 1 million were drawn
for each of the four distributions in Table 5.15, and the empirical estimates
of the cumulants and correlations were computed for both the g-and-h and
GLD transformations. The results are reported in Tables 5.24 and 5.25 for the

TABLE 5.25

The g-and-h (GLD) Empirical Estimates of the Specified
Correlation Matrix in Table 5.16 Using Single Draws of
Size $N = 1,000,000$

	1	2	3	4
1	1			
2	0.300	1		
	(0.299)			
3	0.400	0.600	1	
	(0.400)	(0.600)		
4	0.500	0.699	0.800	1
	(0.501)	(0.702)	(0.801)	

TABLE 5.26

PDFs and Parameters for the Power Method, GLD, and g Distributions

Parameters	Cumulants	PDF
$c_1 = -0.259037$	$\mu = 0$	
$c_2 = 0.867102$	$\sigma^2 = 1$	
$c_3 = 0.265362$	$\gamma_3 = 1.63299$	
$c_4 = 0.021276$	$\gamma_4 = 4$	
$c_5 = 0.002108$	$\gamma_5 = 13.0639$	
$c_6 = 0.000092$	$\gamma_6 = 53.3333$	

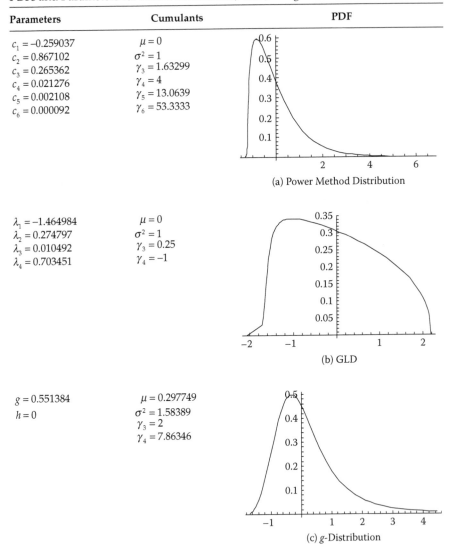

(a) Power Method Distribution

$\lambda_1 = -1.464984$	$\mu = 0$	
$\lambda_2 = 0.274797$	$\sigma^2 = 1$	
$\lambda_3 = 0.010492$	$\gamma_3 = 0.25$	
$\lambda_4 = 0.703451$	$\gamma_4 = -1$	

(b) GLD

$g = 0.551384$	$\mu = 0.297749$	
$h = 0$	$\sigma^2 = 1.58389$	
	$\gamma_3 = 2$	
	$\gamma_4 = 7.86346$	

(c) g-Distribution

TABLE 5.27

Specified Correlation Matrix for the Distributions in Table 5.26

Distribution	Power Method	GLD	g
Power method	1		
GLD	0.4	1	
g	0.5	0.6	1

cumulants and correlations, respectively. Inspection of these tables indicates that the methodology produces estimates of cumulants and correlations that are in close proximity to population parameters.

The g-and-h, GLD, and power method transformations can also be used together for simulating multivariate nonnormal distributions. This may be particularly useful when one transformation has an advantage over another. For example, g transformations can produce *exact* lognormal distributions, whereas the power method or GLD is unable to do so. On the other hand, a fifth-order power method polynomial can match the first six cumulants of a chi-square distribution with degrees of freedom greater than 1, whereas the g-and-h or GLD transformation is unable to do so.

Specifically, consider the three distributions in Table 5.26 with the specified correlations in Table 5.27.

The first distribution is a fifth-order power method pdf with cumulants associated with a chi-square distribution with three degrees of freedom, the second is a GLD pdf with negative kurtosis, and the third is an exact lognormal g distribution. In the context of the GLD, it is noted that neither a (normal-based) polynomial nor a g-and-h transformation can produce a valid pdf for this combination of skew (0.25) and kurtosis (–1). Tables 5.28 and 5.29 give the intermediate correlations and the *Mathematica* source code for determining these correlations.

The empirical estimates of the specified cumulants and correlations are listed in Tables 5.30 and 5.31 and were based on single draws of 1 million pseudorandom deviates. As indicated, the estimates are in close proximity to their respective parameter.

TABLE 5.28

Intermediate Correlations for the Specified Correlation in Table 5.27

Distribution	Power Method	GLD	g
Power method	1		
GLD	0.432142	1	
g	0.535780	0.653415	1

TABLE 5.29

Mathematica Source Code for Solving the Intermediate Correlations in Table 5.28

(* x_1 is a power method polynomial, x_2 is a GLD, and x_3 is a g-and-h distribution. Solved intermediate correlations $\rho_{z_i z_j}$ for the specified correlations in Table 5.27. See Table 5.26 for the parameters of c_i, λ_i, g. *)

$\rho_{z_1 z_2} = 0.432142$

$\rho_{z_1 z_3} = 0.535780$

$\rho_{z_2 z_3} = 0.653415$

$x_1 = c_1 + c_2 z_1 + c_3 z_1^2 + c_4 z_1^3 + c_5 z_1^4 + c z_1^5$

$\phi = \int_{-\infty}^{z_2} \left(\sqrt{2\pi}\right)^{-1} \exp\{-u^2/2\} du$

$x_2 = \lambda_1 + (\phi^{\lambda_3} - (1-\phi)^{\lambda_4})/\lambda_2$

$m = (\exp\{g^2/2\} - 1)/g$

$s = ((\exp\{g^2\} \times (\exp\{g^2\} - 1))/g^2)^{\frac{1}{2}}$

$t = (\exp\{g * z_3\} - 1)/g$

$x_3 = (t - m)/s$

$f_{12} = \left(2\pi\sqrt{1 - \rho_{z_1 z_2}^2}\right)^{-1} \exp\left(-\left(2\left(1 - \rho_{z_1 z_2}^2\right)\right)^{-1}\left(z_1^2 - 2\rho_{z_1 z_2} z_1 z_2 + z_2^2\right)\right)$

$f_{13} = \left(2\pi\sqrt{1 - \rho_{z_1 z_3}^2}\right)^{-1} \exp\left(-\left(2\left(1 - \rho_{z_1 z_3}^2\right)\right)^{-1}\left(z_1^2 - 2\rho_{z_1 z_3} z_1 z_3 + z_3^2\right)\right)$

$f_{23} = \left(2\pi\sqrt{1 - \rho_{z_2 z_3}^2}\right)^{-1} \exp\left(-\left(2\left(1 - \rho_{z_2 z_3}^2\right)\right)^{-1}\left(z_2^2 - 2\rho_{z_2 z_3} z_2 z_3 + z_3^2\right)\right)$

int1 = NIntegrate [$(x_1 x_2) f_{12}$, {Z_1, –8, 8}, {Z_2, –8, 8}, Method → Trapezoidal]
int2 = NIntegrate [$(x_1 x_3) f_{13}$, {Z_1, –8, 8}, {Z_3, –8, 8}, Method → Trapezoidal]
int3 = NIntegrate [$(x_2 x_3) f_{23}$, {Z_2, –8, 8}, {Z_3, –8, 8}, Method → Trapezoidal]

Solution1: int1 = 0.40
Solution2: int2 = 0.50
Solution3: int3 = 0.60

TABLE 5.30

Empirical Estimates of the Specified Correlations in Table 5.27

Distribution	Power Method	GLD	g
Power method	1		
GLD	0.399	1	
g	0.500	0.599	1

Note: The estimates are based on single draws of size 1 million.

TABLE 5.31

Empirical Estimates of the Specified Cumulants in Table 5.26

Distribution	$\hat{\mu}$	$\hat{\sigma}^2$	$\hat{\gamma}_3$	$\hat{\gamma}_4$
Power method	−0.000027	0.999491	1.62717	3.97678
	(0.0)	(1.0)	(1.633)	(4)
GLD	0.000242	0.999649	0.248433	−1.00029
	(0.0)	(1.0)	(0.25)	(−1.0)
g-Distribution	0.298844	1.26007	2.01445	8.02609
	(0.297749)	(1.25853)	(2.0)	(7.863)

Note: The estimates are based on single draws of size 1 million. The parameters are enclosed in parentheses.

Thus, the general procedure for multivariate data generation described here and in previous chapters is flexible to the extent that it allows for not only the simultaneous control of cumulants and correlation structures, but also the type of transformation that is being used, for example, power method polynomial, *g*-and-*h*, or GLD transformation.

References

Badrinath, S. G & Chatterjee, S. (1988). On measuring skewness and elongation in common stock return distributions. The case of the market index. *Journal of Business, 61*, 451–472.

Badrinath, S. G & Chatterjee, S. (1991). A data-analytic look at skewness and elongation in common stock return distributions. *Journal of Business and Economic Statistics, 9*, 223–233.

Bartholomew, D. (1983). Latent variable models for ordered categorical data. *Journal of Econometrics, 22*, 229–243.

Bartko, J. J. (1976). On various intraclass correlation reliability coefficients. *Psychological Bulletin, 83*, 762–765.

Beasley, T. M. (2002). Multivariate aligned rank test for interactions in multiple group repeated measures. *Multivariate Behavioral Research, 37*, 197–226.

Beasley, T. M., Page, G. P., Brand, J. P. L., Gadbury, G. L., Mountz, J. D., & Allison, D. B. (2004). Chebyshev's inequality for nonparametric testing with small N and α in microarray research. *Journal of the Royal Statistical Society, Series C: Applied Statistics, 53*, 95–108.

Beasley, T. M., & Zumbo, B. D. (2003). Comparison of aligned Friedman rank and parametric methods for testing interactions in split-plot designs. *Computational Statistics and Data Analysis, 42*, 569–593.

Benford, F. (1938). The law of anomalous numbers. *Proceedings of the American Philosophical Society, 78*, 551–572.

Besse, P., & Ramsay, J. O. (1986). Principal components analysis of sampled functions. *Psychometrika, 51*, 285–311.

Bickel, P. J., & Doksum, K. A. (1977). *Mathematical Statistics: Basic Ideas and Selected Topics*. Englewood Cliffs, NJ: Prentice-Hall.

Blair, R. C. (1981). A reaction to the "Consequences of failure to meet assumptions underlying the fixed effects analysis of variance and covariance." *Review of Educational Research, 51*, 499–507.

Blair, R. C. (1987). *RANGEN*. Boca Raton, FL: IBM.

Bradley, D. R. (1993). Multivariate simulation with DATASIM: The Mihal and Barrett study. *Behavioral Research Methods, Instruments, and Computers, 25*, 148–163.

Bradley, D. R., & Fleisher, C. L. (1994). Generating multivariate data from nonnormal distributions: Mihal and Barrett revisited. *Behavioral Research Methods, Instruments, and Computers, 26*, 156–166.

Bradley, J. V. (1968). *Distribution Free Statistical Tests*. Englewood Cliffs, NJ: Prentice-Hall.

Bradley, J. V. (1982). The insidious L-shaped distribution. *Bulletin of the Psychonomic Society, 20*, 85–88.

Brunner, E., Domhof, S., & Langer, F. (2002). *Nonparametric Analysis of Longitudinal Data in Factorial Experiments*. New York: John Wiley.

Christian, J. C., Yu, P. L., Slemenda, C. W., & Johnston, C. C., Jr. (1989). Heritability of bone mass: A longitudinal study in aging male twins. *American Journal of Human Genetics, 44*, 429–433.

Conover, W. J., & Iman, R. L. (1981). Rank transformations as a bridge between parametric and nonparametric statistics. *American Statistician, 35,* 124–133.

Cook, R. D., & Weisberg, S. (1999). *Applied Regression Including Computing and Graphics.* New York: John Wiley.

Corrado, C. J. (2001). Option pricing based on the generalized lambda distribution. *Journal of Future Markets, 21,* 213–236.

Cronbach, L. J. (1951). Coefficient alpha and the internal structure of tests. *Psychometrika, 16,* 297–334.

Cronbach, L. J., Gleser, G. C., Nanda, H., & Rajaratnam, N. (1972). *The Dependability of Behavioral Instruments: Theory of Generalizability for Scores and Profiles.* New York: Wiley.

Delaney, H. D., & Vargha, A. (2000). The *Effect on Non-Normality on Student's Two Sample T-Test.* Paper presented at the annual meeting of the American Educational Research Association, New Orleans.

Demirtas, H., & Hedeker, D. (2008). Multiple imputation under power polynomials. *Communications in Statistics—Simulation and Computation, 37,* 1682–1695.

Devroye, L. (1986). *Non-Uniform Random Variate Generation.* New York: Springer.

Dudewicz, E. J., & Karian, Z. A. (1996). The EGLD (extended generalized lambda distribution) system for fitting distributions to data with moments. II. Tables. *American Journal of Mathematical and Management Sciences, 16,* 271–332.

Dudewicz, E. J., & Karian, Z. A. (1999). The role of statistics in IS/IT: Practical gains from mined data. *Information Systems Frontiers, 1,* 259–266.

Dupuis, D. J., & Field, C. A. (2004). Large windspeeds, modeling and outlier detection. *Journal of Agricultural, Biological and Environmental Statistics, 9,* 105–121.

Dutta, K. K., & Babbel, D. F. (2005). Extracting probabilistic information from the prices of interest rate options: Tests of distributional assumptions. *Journal of Business, 78,* 841–870.

Dwivedi, T., & Srivastava, K. (1978). Optimality of least squares in the seemingly unrelated regression model. *Journal of Economics, 7,* 391–395.

Field, C. A., & Genton, M. G. (2006). The multivariate *g*-and-*h* distribution. *Technometrics, 48,* 104–111.

Finch, H. (2005). Comparison of the performance of nonparametric and parametric MANOVA test statistics when assumptions are violated. *Methodology, 1,* 27–38.

Fisher, R. A. (1921). On the probable error of a coefficient of correlation deduced from a small sample. *Metron, 1,* 1–32.

Fleishman, A. I. (1978). A method for simulating non-normal distributions. *Psychometrika, 43,* 521–532.

Freimer, M., Mudholkar, G. S., Kollia, G., & Lin, T. (1988). A study of the generalized Tukey lambda family. *Communications in Statistics: Theory and Methods, 17,* 3547–3567.

Ganeshan, R. (2001). Are more suppliers better? Generalizing the Gau and Ganeshan procedure. *Journal of the Operations Research Society, 52,* 122–123.

Gibbons, J. D., & Chakraborti, S. (1992). *Nonparametric Statistical Inference* (3rd ed.). New York: Marcel Dekker.

Gini, C. (1912). Variabilita' e Mutabilita', contributo allo studio delle distribuzioni e relazioni statistiche. *Studi Econonmico-Giuridici dell' Universita' di Cagliari, 3,* 1–158.

Giraudeau, B., & Mary, J. Y. (2001). Planning a reproducibility study: How many subjects and how many replicates per subject for an expected width of the 95 per cent confidence interval of the intraclass correlation coefficient. *Statistics in Medicine, 20*, 3205–3214.

Green, W. H. (1993). *Econometric Analysis* (2nd ed.). New York: McMillan.

Habib, A. R., & Harwell, M. R. (1989). An empirical study of the type I error rate and power of some selected normal theory and nonparametric tests of independence of two sets of variables. *Communications in Statistics: Simulation and Computation, 18*, 793–826.

Hanisch, D., Dittmar, M., Hohler, T., & Alt, K. W. (2004). Contribution of genetic and environmental factors to variation in body compartments—a twin study in adults. *Anthropologischer Anzeiger, 62*, 51–60.

Harwell, M. R., & Serlin, R. C. (1988). An experimental study of a proposed test of nonparametric analysis of covariance. *Psychological Bulletin, 104*, 268–281.

Harwell, M. R., & Serlin, R. C. (1989). A nonparametric test statistic for the general linear model. *Journal of Educational Statistics, 14*, 351–371.

Harwell, M. R., & Serlin, R. C. (1997). An empirical study of five multivariate tests for the single-factor repeated measures model. *Communications in Statistics: Simulation and Computation, 26*, 605–618.

Headrick, T. C. (1997). *Type I Error and Power of the Rank Transform Analysis of Covariance (ANCOVA) in a 3 × 4 Factorial Layout.* Unpublished doctoral dissertation, Wayne State University, Detroit, MI.

Headrick, T. C. (2002). Fast fifth-order polynomial transforms for generating univariate and multivariate non-normal distributions. *Computational Statistics and Data Analysis, 40*, 685–711.

Headrick, T. C. (2004). On polynomial transformations for simulating multivariate non-normal distributions. *Journal of Modern Applied Statistical Methods, 3*, 65–71.

Headrick, T. C., Aman, S. Y., & Beasley, T. M. (2008a). A method for simulating correlated structures of continuous and ranked data. *Communications in Statistics: Simulation and Computation, 37*, 602–616.

Headrick, T. C., & Beasley, T. M. (2004). A method for simulating correlated non-normal systems of linear statistical equations. *Communications in Statistics: Simulation and Computation, 33*, 19–33.

Headrick, T. C., & Kowalchuck, R. K. (2007). The power method transformation: Its probability density function, distribution function, and its further use for fitting data. *Journal of Statistical Computation and Simulation, 77*, 229–249.

Headrick, T. C., Kowalchuck, R. K., & Sheng, Y. (2008b). Parametric probability densities and distribution functions for Tukey g-and-h transformation and their use for fitting data. *Applied Mathematical Sciences, 2*, 449–462.

Headrick, T. C., & Mugdadi, A. (2006). On simulating multivariate non-normal distributions from the generalized lambda distribution. *Computational Statistics and Data Analysis, 50*, 3343–3353.

Headrick, T. C., & Rotou, O. (2001). An investigation of the rank transformation in multiple regression. *Computational Statistics and Data Analysis, 38*, 203–215.

Headrick, T. C., & Sawilowsky, S. S. (1999a). Simulating correlated non-normal distributions: Extending the Fleishman power method. *Psychometrika, 64*, 25–35.

Headrick, T. C., & Sawilowsky, S. S. (1999b, January). *The Best Test for Interaction in Factorial ANOVA and ANCOVA*. University of Florida Statistics Symposium on Selected Topics in Nonparametric Methods, Gainesville, FL.

Headrick, T. C., & Sawilowsky, S. S. (2000a). Properties of the rank transformation in factorial analysis of covariance. *Communications in Statistics: Simulation and Computation, 29*, 1059–1088.

Headrick, T. C., & Sawilowsky, S. S. (2000b). Weighted simplex procedures for determining boundary points and constants for the univariate and multivariate power methods. *Journal of Educational and Behavioral Statistics, 25*, 417–436.

Headrick, T. C., Sheng, Y., & Hodis, F. A. (2007). Numerical computing and graphics for the power method transformation using *Mathematica. Journal of Statistical Software, 19*, 1–17.

Headrick, T. C., & Vineyard, G. (2001). An empirical investigation of four tests for interaction in the context of factorial analysis of covariance. *Multiple Linear Regression Viewpoints, 27*, 3–15.

Headrick, T. C., & Zumbo, B. D. (2008). A method for simulating multivariate non-normal distributions with specified standardized cumulants and intraclass correlation coefficients. *Communications in Statistics: Simulation and Computation, 37*, 617–628.

Hendrix, L., & Habing, B. (2009, April). *MCMC Estimation of the 3PL Model Using a Multivariate Prior Distribution*. Paper presented at the annual meeting of the American Educational Research Association, San Diego.

Hess, B., Olejnik, S., & Huberty, C. J. (2001). The efficacy of two improvement-over-chance effect sizes for two group univariate comparisons under variance heterogeneity and nonnormality. *Educational and Psychological Measurement, 61*, 909–936.

Hipp, J. R., & Bollen, K. A. (2003). Model fit in structural equation models with censored, ordinal, and dichotomous variables: Testing vanishing tetrads. *Sociological Methodology, 33*, 267–305.

Hoaglin, D. C. (1983). *g*-and-*h* distributions. In S. Kotz & N. L. Johnson (Eds.), *Encyclopedia of Statistical Sciences* (Vol. 3, pp. 298–301). New York: Wiley.

Hoaglin, D. C. (1985). Summarizing shape numerically: The *g*-and-*h* distributions. In D. C. Hoaglin, F. Mosteller, & J. W. Tukey (Eds.), *Exploring Data, Tables, Trends, and Shapes* (pp. 461–511). New York: Wiley.

Hodis, F. A. (2008). *Simulating Univariate and Multivariate Non-Normal Distributions Based on a System of Power Method Distributions*. Unpublished doctoral dissertation, Southern Illinois University, Carbondale.

Hodis, F. A., & Headrick, T. C. (October, 2007). *A System of Power Method Distributions Based on the Logistic, Normal, and Uniform Distributions*. Paper presented at annual meeting of the Mid-Western Educational Research Association, St. Louis, MO.

Hotelling, H. (1953). New light on the correlation coefficient and its transforms. *Journal of the Royal Statistical Society (Series B), 2*, 193–224.

Johnson, N. L., Kotz, S., & Balakrishnan, N. (1994). *Continuous Univariate Distributions* (Vol. 1, 2nd ed.). New York: John Wiley.

Judge, G. J., Hill, R. C., Griffiths, W. E., Lutkepohl, H., & Lee, T. C. (1985). *Introduction to the Theory and Practice of Econometrics* (2nd ed.). New York: John Wiley.

Karvanen, J. (2003). *Generation of Controlled Non-Gaussian Random Variables from Independent Components.* Paper presented at the Fourth International Symposium of Independent Component Analysis and Blind Signal Separation, Nara, Japan.

Karian, Z. A., & Dudewicz, E. J. (2000). *Fitting Statistical Distributions: The Generalized Lambda Distribution and Generalized Bootstrap Methods.* Boca Raton, FL: Chapman & Hall/CRC.

Karian, Z. A., Dudewicz, E. J., & McDonald, P. (1996). The extended generalized lambda distribution system for fitting distributions to data: History, completion of theory, tables applications, the "final word" on moments fits. *Communications in Statistics: Simulation and Computation, 25,* 611–642.

Kendall, M. G., & Gibbons, J. D. (1990). *Rank Correlation Methods* (5th ed.). New York: Oxford.

Kendall, M., & Stuart, A. (1977). *The Advanced Theory of Statistics* (4th ed.). New York: Macmillan.

Keselman, H. J., Kowalchuk, R. K., & Lix, L. M. (1998). Robust nonorthogonal analysis revisited: An update on trimmed means. *Psychometrika, 63,* 145–163.

Keselman, H. J., Lix, L. M., & Kowalchuk, R. K. (1998). Multiple comparison procedures for trimmed means. *Psychological Methods, 3,* 123–141.

Keselman, H. J., Wilcox, R. R., Kowalchuk, R. K., & Olejnik, S. (2002). Comparing trimmed or least squares means of two independent skewed populations. *Biometrical Journal, 44,* 478–489.

Keselman, H. J., Wilcox, R. R., Taylor, J., & Kowalchuk, R. K. (2000). Tests for mean equality that do not require homogeneity of variances: Do they really work? *Communications in Statistics: Simulation and Computation, 29,* 875–895.

Klockars, A. J., & Moses, T. P. (2002). Type I error rates for rank-based tests of homogeneity of regression slopes. *Journal of Modern Applied Statistical Methods, 1,* 452–460.

Kotz, S., Balakrishnan, N., & Johnson, N. L. (2000). *Continuous Multivariate Distributions* (2nd ed.). New York: John Wiley.

Kowalchuk, R. K., & Headrick, T. C. (2008, March). *Simulating Correlated Multivariate Non-Normal g-and-h Distributions.* Paper presented at the annual meeting of the American Educational Research Association, New York.

Kowalchuk, R. K., & Headrick, T. C. (2009). Simulating multivariate g-and-h distributions. *British Journal of Mathematical and Statistical Psychology,* in press, doi: 10.1348/000711009X423067.

Kowalchuk, R. K., Keselman, H. J., & Algina, J. (2003) Repeated measures interaction test with aligned ranks. *Multivariate Behavioral Research, 38,* 433–461.

Kowalchuk, R. K., Keselman, H. J., Wilcox, R. R., & Algina, J. (2006). Multiple comparison procedures, trimmed means and transformed statistics. *Journal of Modern Applied Statistical Methods, 5,* 44–65.

Lix, L. M., Algina, J., & Keselman, H. J. (2003). Analyzing multivariate repeated measures designs: A comparison of two approximate degrees of freedom procedures. *Multivariate Behavioral Research, 38,* 403–431.

Marsaglia, G. (2004). Evaluating the normal distribution. *Journal of Statistical Software, 11,* 1–11.

Martinez, J., & Iglewicz, B. (1984). Some properties of the Tukey g-and-h family of distributions. *Communications in Statistics: Theory and Methods, 13,* 353–369.

McGraw, K. O., & Wong, S. P. (1996). Forming inferences about some intraclass correlation coefficients. *Psychological Methods, 1,* 30–46.

Micceri, T. (1989). The unicorn, the normal curve and other improbable creatures. *Psychological Bulletin, 105,* 156–166.

Mills, T. C. (1995). Modelling skewness and kurtosis in the London Stock Exchange FT-SE index return distributions. *Journal of the Royal Statistical Society: Series D, 44,* 323–332.

Moran, P. A. P. (1948). Rank correlation and product-moment correlation. *Biometrika, 35,* 203–206.

Morgenthaler, S., & Tukey, J. W. (2000). Fitting quantiles: Doubling, HR, HQ, and HHH distributions. *Journal of Computational and Graphical Statistics, 9,* 180–195.

Murray, D. M., Rooney, B. L., Hannan, P. J., Peterson, A. V., Ary, D. V., Biglan, A., et al. (1994). Intraclass correlation among common measures of adolescent smoking: Estimates, correlates, and applications in smoking preventions studies. *American Journal of Epidemiology, 140,* 1038–1050.

Mutihac, R., & Van Hulle, M. M. (2003). A comparative study on adaptive neural network algorithms for independent component analysis. *Romanian Reports in Physics, 55,* 43–47.

Neter, J., Kutner, M. H., Nachtsheim, C. J., & Wasserman, W. (1996). *Applied Linear Statistical Models* (4th ed.). Boston: WCB/McGraw-Hill.

Olejnik, S. F., & Algina, J. (1987). An analysis of statistical power for parametric ANCOVA and rank transform ANCOVA. *Communications in Statistics: Theory and Methods, 16,* 1923–1949.

Pearson, E. S., & Please, N. W. (1975). Relationship between the shape of population distribution and robustness or four simple testing statistics. *Biometrika, 63,* 223–241.

Pearson, K. (1907). *Mathematical Contributions to the Theory of Evolution. XVI. On Further Methods of Determining Correlation.* Draper's Research Memoirs, Biometric Series IV, Cambridge University Press.

Powell, D. A., Anderson, L. M., Chen, R. Y. S., & Alvord, W. G. (2002). Robustness of the Chen-Dougherty-Bittner procedure against non-normality and heterogeneity in the coefficient of variation. *Journal of Biomedical Optics, 7,* 650–660.

Ramberg, J. S., Dudewicz, E. J., Tadikamalla, P. R., & Mykytka, E. F. (1979). A probability distribution and its uses in fitting data. *Technometrics, 21,* 201–214.

Ramberg, J. S., & Schmeiser, B. W. (1972). An approximate method for generating symmetric random variables. *Communications of the ACM, 15,* 987–990.

Ramberg, J. S., & Schmeiser, B. W. (1974). An approximate method for generating asymmetric random variables. *Communications of the ACM, 17,* 78–82.

Rasch, D., & Guiard, V. (2004). The robustness of parametric statistical methods. *Psychology Science, 46,* 175–208.

Reinartz, W. J., Echambadi, R., & Chin, W. W. (2002). Generating non-normal data for simulation of structural equation models using Mattson's method. *Multivariate Behavioral Research, 37,* 227–244.

Rowe, A. K., Lama, M., Onikpo, F., & Deming, M. S. (2002). Design effects and intraclass correlation coefficients from a health facility cluster survey in Benin. *International Journal of Quality in Health Care, 14,* 521–523.

Sawilowsky, S. S., & Blair, R. C. (1992). A more realistic look at the robustness of the type II error properties of the t-test to departures from population normality. *Psychological Bulletin, 111,* 352–360.

Scheffe, H. (1959). *The Analysis of Variance.* New York: John Wiley & Sons.

Serlin, R.C., & Harwell, M.A. (2004). More powerful tests of predictor subsets in regression analysis under nonnormality. *Psychological Methods, 9*, 492–509.

Shieh, Y. (April, 2000). *The Effects of Distributional Characteristics on Multi-Level Modeling Parameter Estimates and Type I Error Control of Parameter Tests under Conditions Of Non-Normality*. Paper presented at the annual meeting of the American Educational Research Association, New Orleans.

Shrout, P. E., & Fleiss, J. L. (1979). Intraclass correlations: Uses in assessing rater reliability. *Psychological Bulletin, 86*, 420–428.

Siddiqui, O., Hedeker, D., Flay, B. R., & Hu, F. B., (1996). Intraclass correlation estimates in a school-based smoking prevention study. Outcome and mediating variables, by sex and ethnicity. *American Journal of Epidemiology, 144*, 424–433.

S-Plus 8.0.4 for Windows. (2007). Palo Alto, CA: TIBCO Software.

Stevenson, W. R., & Jacobson, D. (1988). A comparison of non-parametric analysis of covariance techniques. *Communications in Statistics: Simulation and Computation, 26*, 605–618.

Steyn, H. S. (1993). On the problem of more than one kurtosis parameter in multivariate analysis. *Journal of Multivariate Analysis, 44*, 1–22.

Stone, C. (2003). Empirical power and type I error rates for an IRT fit statistic that considers the precision and ability estimates. *Educational and Psychological Measurement, 63*, 566–583.

Stuart, A. (1954). The correlation between variate-values and ranks in samples from a continuous distribution. *British Journal of Statistical Psychology, 7*, 37–44.

Tadikamalla, P. R. (1980). On simulating nonnormal distributions. *Psychometrika, 45*, 273–279.

Tang, X., & Wu, X. (2006). A new method for the decomposition of portfolio VaR. *Journal of Systems Science and Information, 4*, 721–727.

Tukey, J. W. (1960). *The Practical Relationship between the Common Transformation of Percentages of Counts and of Amounts*. Technical Report 36, Statistical Techniques Research Group, Princeton University.

Tukey, J. W. (1977). *Modern Techniques in Data Analysis*. NSF-sponsored regional research conference at Southern Massachusetts University, North Darmouth.

Vale, C. D., & Maurelli, V. A. (1983). Simulating multivariate nonnormal distributions. *Psychometrika, 48*, 465–471.

Vargha, A., & Delaney, H. D. (1998). The Kruskal-Wallis test and stochastic homogeneity. *Journal of Educational and Behavioral Statistics, 23*, 170–192.

Vargha, A., & Delaney, H. D. (2000). A critique and improvement of the CL common language effect size statistic of McGraw and Wong. *Journal of Educational and Behavioral Statistics, 25*, 101–132.

Welch, G., & Kim, K. H. (2004, May). *An Evaluation of the Fleishman Transformation for Simulating Non-Normal Data in Structural Equation Modeling*. Paper presented at the international meeting of the Psychometric Society, Monterey, CA.

Wilcox, R. R. (2001). Detecting nonlinear associations, plus comments on testing hypothesis about the correlation coefficient. *Journal of Educational and Behavioral Statistics, 26*, 73–83.

Wilcox, R. R. (2006). Inferences about the components of a generalized additive model. *Journal of Modern Applied Statistical Methods, 5*, 309–316.

Wilcox, R. R., Keselman, H. J., & Kowalchuk, R. K. (1998). Can tests for treatment group equality be improved? The bootstrap and trimmed means conjecture. *British Journal of Mathematical and Statistical Psychology, 51*, 123–134.

Wolfram, S. (2003). *The* Mathematica *Book* (5th ed.). Champaign, IL: Wolfram Media, Inc.

Wong, S. P., & McGraw, K. O. (2005). Confidence intervals and F tests for intraclass correlations based on three-way random effects models. *Educational and Psychological Measurement, 59*, 270–288.

Zellner, A. (1962). An efficient method of estimating seemingly unrelated regressions and tests of aggregation bias. *Journal of the American Statistical Association, 57*, 53–73.

Zhu, R., Yu, F., & Liu, S. (2002, April). *Statistical Indexes for Monitoring Item Behavior under Computer Adaptive Testing Environment*. Paper presented at the annual meeting of the American Educational Research Association, New Orleans.

Index